T0210644

Wireless Networks

Series Editors
Xuemin (Sherman) Shen

More information about this series at http://www.springer.com/series/14180

Kuan Zhang • Xuemin (Sherman) Shen

Security and Privacy for Mobile Healthcare Networks

 Springer

Kuan Zhang
University of Waterloo
Waterloo, ON, Canada

Xuemin (Sherman) Shen
University of Waterloo
Waterloo, ON, Canada

Wireless Networks
ISBN 978-3-319-36997-6   ISBN 978-3-319-24717-5 (eBook)
DOI 10.1007/978-3-319-24717-5

Printed on acid-free paper

Springer International Publishing AG Switzerland is part of Springer Science+Business Media (www.springer.com)

Preface

Healthcare has been a critical social and economic issue worldwide, especially in the aging society, with tremendous health expenses and labor resources. Furthermore, the traditional hospital-centric healthcare system lacks efficiency when dealing with chronic diseases or identifying some serious diseases in the early stage. It also suffers from excessive waiting time for patients in hospital. Therefore, developing a continuous and ubiquitous health monitoring system becomes a promising approach to offer pervasive healthcare services.

With the widespread use of smartphones and the advance of versatile wearable wireless devices, mobile healthcare network (MHN) becomes a prestigious paradigm of ubiquitous healthcare to continuously monitor people's health conditions, remotely diagnose phenomena, and share health information in a real-time pattern. However, security and privacy concerns become critical, since MHNs collect personal private health information, and users have various security requirements and privacy levels in different applications.

In this monograph, we investigate security and privacy issues in MHNs and provide adjustable security and privacy protections. Firstly, we identify the overall architecture of MHNs, including wearable devices, mobile users, heterogeneous network, and cloud servers. Based on the MHN architecture, we present several emerging health-related applications, i.e., clinic/remote healthcare, home care, fitness and training, as well as some social networking applications. Secondly, we investigate general security and privacy requirements in MHNs and identify related challenges, such as privacy leakage, access control, privacy-preserving health data processing, and misbehaviors in MHNs. These security and privacy challenges embrace the emerging MHN applications and compromise the traditional security and privacy preservation techniques. To address these challenging issues, we propose effective security and privacy protection schemes in MHNs as follows.

Chapter 3 proposes a priority based health data aggregation (PHDA) scheme with privacy preservation for cloud-assisted WBANs to address the security and privacy problems during health data transmission in MHNs. The proposed PHDA scheme reduces the aggregation overheads and preserves user privacy at the same time. The health data are categorized into several types, and each type is assigned with a specific priority. When users upload their data, they can select different forwarding strategies according to the health data priority. During the aggregation phase, the health data with higher priority can be forwarded within a smaller delay. In addition, the data with the same priority can be aggregated efficiently to reduce the communication overheads.

Chapter 4 presents a social-based mobile Sybil detection (SMSD) scheme to address the misbehaving issues during information sharing in MHNs. The SMSD scheme detects Sybil attackers according to their abnormal contacts and pseudonym-changing behaviors in mobile networks. Since the Sybil attackers frequently change their pseudonyms to cheat other users, we investigate the contact statistics of the used pseudonyms and detect Sybil attackers by comparing the contact statistics of pseudonyms from normal users and that from Sybil attackers. The cloud servers are also adopted to store and process the large volume of user's contact information, alleviating the burden of mobile users. The SMSD also addresses the collusion attacks and resists cloud data modification when employing the cloud server for mobile Sybil detection.

Chapter 5 investigates privacy-preserving health data processing schemes. We introduce several basic classification schemes and homomorphic encryptions as preliminaries. Then, we present the integration of classification schemes and secure computation blocks. Finally, we discuss an application of clinic decision support system and provide privacy preservation solutions.

Chapter 6 presents some emerging attribute-based access control (ABAC) schemes for health data access control in MHNs. We investigate some popular ABE schemes, such as CP-ABE, KP-ABE, and multi-authority-based ABE schemes. Based on these techniques, we provide the integration of ABE schemes to achieve access control in MHNs.

Finally, we summarize the monograph and point out some open problems for future research directions in MHNs.

Waterloo, Canada   Kuan Zhang
                                                                          Xuemin (Sherman) Shen

Acknowledgements

We would like to thank Prof. Rongxing Lu from Nanyang Technology University, Singapore; Prof. Xiaodong Lin from University of Ontario Institute and Technology, Oshawa, ON, Canada; Prof. Zhou Su from Shanghai University, Shanghai, China; Prof. Juntao Gao from Xidian University, Xi'an, China; Dr. Henry. H. Luo from Care-In-Motion Technology (CIM) Inc., Kitchener, ON, Canada; Dr. Xiaohui Liang from Dartmouth College, Hanover, NH, USA; and Dr. Kan Yang, Dr. Mrinmoy Barua, Dr. Ning Lu, Dr. Ning Zhang, Mr. Ju Ren, Mr. Jianbing Ni, and Mr. Nan Cheng from Broadband Communications Research (BBCR) Group at the University of Waterloo, Waterloo, ON, Canada, for their contributions in the presented research works. We also would like to thank all the members of BBCR group for the valuable discussions and their insightful suggestions, ideas, and comments. Special thanks are also due to the staff at Springer Science+Business Media, Susan Lagerstrom-Fife and Jennifer Malat, for their help throughout the publication preparation process.

Contents

Acronyms

ABE	Attribute-Based Encryption
AP	Access Point
BCSD	Behavior Classification-based Sybil Detection
BS	Base Station
BT	Bluetooth
CP-ABE	Ciphertext Policy Attribute-Based Encryption
CS	Cloud Server
D2D	Device-to-Device
DBDH	Decisional Bilinear Diffie-Hellman
DDH	Decisional Diffie-Hellman
DMBDH	Decisional Modified Bilinear Diffie-Hellman
DoS	Denial-of-Service
DSRC	Dedicated Short-Range Communications
ECG	Electrocardiography
EM	Expectation-Maximization
FHE	Fully Homomorphic Encryption
FNR	False Negative Rate
FPR	False Positive Rate
FR-MSD	Friend Relationship-based Mobile Sybil Detection
GMM	Gaussian Mixture Model
GPS	Global Positioning System
HE	Homomorphic Encryption
HMM	Hidden Markov Model
KCRFs	Kernel Conditional Random Fields
KP-ABE	Key Policy Attribute-Based Encryption
kPC	k-anonymity-based Pseudonym Changing
LSMSD	Learning-assisted SMSD
LSP	Local Service Provider
LTE	Long-Term Evolution
MA-ABE	Multi-Authority Attribute-Based Encryption
MAC	Message Authentication Code

MAP	Maximum A Posteriori
MHN	Mobile Healthcare Network
MSD	Mobile Sybil Detection
MSN	Mobile Social Network
OSN	Online Social Network
PDF	Probability Density Function
PHDA	Priority-based Health Data Aggregation
PKG	Public Key Generator
PBPC	Period-Based Pseudonym Changing
PPAM	Privacy-Preserving ArgMax
PPDP	Privacy-Preserving Dot Product
PPC	Privacy-Preserving Comparison
QoS	Quality of Service
QoE	Quality of Experience
QoP	Quality of Protection
QR	Quadratic Residuosity
RSS	Received Signal Strength
RSU	Roadside Unit
SA	Sybil Attack
SC	Social Community
SCSD	Social Community-based Sybil Detection
SMSD	Social-based Mobile Sybil Detection
SNSD	Social Network-based Sybil Detection
SP	Social Proximity
SS	Social Spot
SVM	Support Vector Machine
SWHE	SomeWhat Homomorphic Encryption
TA	Trusted Authority
VANET	Vehicular Ad Hoc NETwork
WBAN	Wireless Body Area Network

Chapter 1
Introduction

Healthcare becomes one of major economic and social problems around the world, especially in the aging societies, where it costs tremendous health expenses and human resources [1]. According to a recent national health report of the United States, an average expense per capita is $8,895 in 2014. The annual national healthcare expenditure skyrockets to $3.8 trillion, where nursing care, home healthcare and personal care account for about 18 % [2]. Furthermore, the ever-increasing aging population in the worldwide poses new challenges for healthcare. In 2010, the population of people aged 65 or older was 524 million, which represents 8 % of the world's population. By 2050, the aging population is expected to about 2 billion, representing 16 % of the world overall population. In addition, senior people likely suffer from chronic diseases, such as heart and respiratory diseases. Chronic diseases are also poorly treated in current hospital-centric healthcare system. This traditional hospital-centric healthcare system not only lacks efficiency when identifying some serious diseases and dealing with chronic diseases in the early stage, but also suffers from excessive waiting time in hospital. Thus, it is necessary to pose up-and-coming healthcare services, such as continuous health monitoring, health information processing and sharing, to enhance the early disease diagnosis and relieve the heavy burden of the current health expenditure.

1.1 Overview of Mobile Healthcare Networks

Recently, wearable devices, such as smart wristwatch, ring, bracelet and hairlace, are widely applied to offer continuous healthcare, e.g., physiology parameter monitoring for remote healthcare [3], heart rate record [4, 5] for workout intensity or training, and calorie burn during fitness. These smartwatches, health monitors, pedometers, activity trackers and virtual reality headsets are all part of the emerging landscape of wearable technology, which promises to not only change the way we

© Springer International Publishing Switzerland 2015
K. Zhang, X. Shen, *Security and Privacy for Mobile Healthcare Networks*, Wireless Networks, DOI 10.1007/978-3-319-24717-5_1

exercise and communicate but also support the emerging healthcare. As wearable technologies are rapidly evolving and the market is poised to explode, the demands for wearable devices are forecasted to generate $53.2 billion by 2019 [6]. Consisting of these ubiquitous wearable devices, mobile healthcare networks (MHNs) take the advantages of heterogeneous mobile networks (e.g., cellular network, WiFi and Device-to-Device (D2D) communications) and powerful computational servers (e.g., cloud server) to collect the health information sensed by wearable devices, analyze/process the information for health monitoring and diagnosis, and enable users' social interactions [1, 7]. For example, the seniors can wear the dedicated wearable devices which continuously measure the seniors' physiology information, such as body temperature, heart rate, blood pressure and Oxygen saturation. Meanwhile, doctors or seniors' families can use desktops and smartphones to remotely access these health records via MHNs. In case of any emergency, such as falling down and heart problem, the wearable devices can automatically report the health condition of the patient to his doctors and families via MHNs. In addition, MHNs can also enable promising wearable and social applications [8], e.g., sharing physical condition and activity information measured by wearable devices among social friends [9, 10].

1.2 MHN Architecture and Applications

In this section, we propose the architecture of MHN, where different communication and computation functionalities of each component are provided. Then, we introduce some promising MHN applications under the proposed MHN architecture.

1.2.1 MHN Architecture

In MHNs, users' health-related data should be measured, transmitted, analyzed and processed to support different applications. To achieve these functionalities, an MHN consists of wearable devices, users, servers and heterogenous mobile networks as shown in Fig. 1.1.

1.2.1.1 Wearable Devices

Wearable devices, as the bridge connecting the human body and information world, are integrated with physiology sensors, low-power computation, communication and storage modules. These devices can sense diverse information from human, such as physiology parameters, health condition, motions and location. Generally, wearable devices can only pre-process the sensed data due to the limitations of size, processing capabilities and energy. Alternatively, these sensed data are compressed

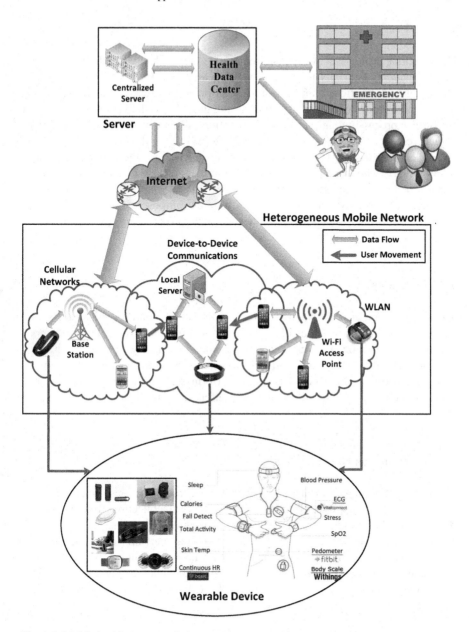

Fig. 1.1 Mobile healthcare network

by the embedded low-power computation modules, sent to mobile users' devices (i.e., smartphones and desktops) via Bluetooth or NFC, or directly delivered to the servers via heterogeneous mobile networks. The communications among wearable devices is also based on Bluetooth, NFC, red infrared communication, etc.

1.2.1.2   Users

Users include health data owners and requestors. On one hand, health data owners, such as patients or MHN users, take various wearable devices to measure their physiology information. They also have smartphones to receive the sensing data from wearable devices. These measured health data can be delivered to the servers for further processing and analysis. Furthermore, users can be either the sensing objects (e.g., patients and seniors) of wearable devices, or the monitor to measure and collect the health data from the sensing objects. On the other hand, health data requestors, e.g., hospital, doctors, patient's families, insurance company, etc., can request the collected health data from data owners and analyze them to obtain the desired results from MHNs. According to different purposes of health data usage and roles of different data requestors, the authorization and access policies for data requestors are also different.

1.2.1.3   Servers

The servers, such as centralized servers in hospital and cloud servers, are used to store, process and analyze the collected health data from the wearable devices and data owners. Centralized servers store a large amount of data owner's health data and have powerful processing capabilities. Data requestors access health data via these servers. In addition, some local servers can perform as an authority to automatically organize the local MHNs and provide local information to facilitate mobile users' interactions.

1.2.1.4   Heterogeneous Mobile Networks

Consisting of cellular network, WiFi, and D2D communications, heterogeneous mobile networks support MHNs for the health data collection from wearable devices and data owners, transmission and sharing among users. MHNs can be seamlessly switched among different types of wireless networks during the health data collection, transmission and sharing. With heterogeneous mobile networks, mobile users can access the Internet through WiFi or cellular network, interact with surrounding users via Bluetooth or NFC, and browse the local information via local servers.

1.2.2   MHN Applications

Under the above MHN architecture, there are various MHN applications, including clinic healthcare, home care and fitness, as shown in Fig. 1.2.

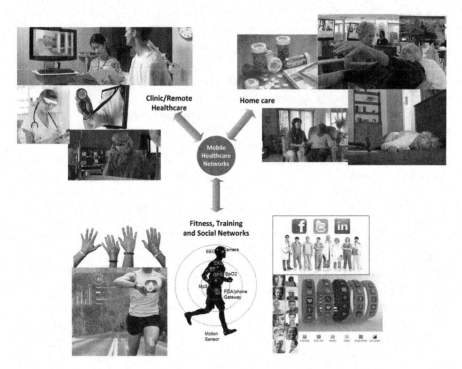

Fig. 1.2   MHN applications

1.2.2.1   Clinic Healthcare

Clinic healthcare is one of the most prestigious MHN applications, which can offer continuous physiology parameter sensing and health condition monitoring. In the hospital, the multi-functional wearable devices can measure patient's physiology and activity parameters, such as heart rate, body temperature, blood pressure, Oxygen saturation, blood volume index, respiration, pulse, quality of sleep, location, fall and posture. Based on these measured health information of patient, doctors and related family members can remotely and real-timely monitor the patient's health condition, such as blood pressure to identify hypertension, Glucagon to indicate diabetes, Oxygen saturation to diagnose myocardial infarction (MI) or acute myocardial infarction (AMI), etc.

Clinic healthcare provides health monitoring for the patients in hospital environments. The measured data are mainly used for clinic diagnosis and monitoring. The measurement devices can be wearable devices (e.g., bracelet and headsets) or dedicated medical equipments. The transmission reliability should be guaranteed, and delay is usually low. The analysis of the measured physiology and activity parameters should also be performed within a short period. The accuracy of the measurement and analysis should be very high. Warning or alarm system is associated with clinic healthcare applications for emergency cases.

1.2.2.2   Home Care

Home care can offer the ubiquitous healthcare for the seniors, disabled people, even though they are staying at home. Home care considerably saves the resources of hospital, nursing homes, aged persons homes, etc. It also makes the health monitoring convenient for seniors and disabled people.

One promising home care application is fall detection, which offers emergency response with MHNs. In fall detection application, the abnormal body position can trigger the acceleration sensors worn on the body of seniors and disabled people to identify the fall. After detecting people's fall, wearable devices or smartphones can report this emergency to the patient's family and doctors through MHNs in a real time pattern. Furthermore, the real-time physiology parameters of falling person can be measured and transmitted to hospital and doctors, offering a guideline for the emergency operation. The embedded GPS module can report the patient's location associated with the alarm message to hospital.

Another typical home care application is the early detection of chronic disease and prediction of sudden disease. Some critical diseases, such as myocardial infarction, heart attack and stroke, may be reflected by certain abnormal body conditions, e.g., ever-decreasing Oxygen saturation and high pulse. However, without MHNs, these symptoms are hardly identified by patients. Home care applications take the advantages of measurement, transmission and analysis with MHNs to provide health condition prediction and chronic disease detection in the early stage, which may save people's life with a higher probability.

Home care applications are usually used in non-hospital environments. The results of home care applications can be the guidelines or assistances for disease prevention and doctor's further diagnosis. The analysis accuracy can be within a reasonable range. The transmission delay varies with different home care applications.

1.2.2.3   Fitness and Training

Despite the aforementioned health monitoring related applications, MHNs can also offer a wide range of applications, including fitness and training. The wearable devices (e.g., belt, glove and bracelet) are able to capture the motion of the human body, arms or hands, which captures calorie burn, heart or lung conditions during fitness and training. The sensed health data would become the main driver of users' further fitness plan or the coach's decisions. During the fitness, users can also share the physiology parameters with other users for experience sharing or feedback [11]. For example, they can share the suitable fitness guideline to the users with the similar body conditions, or recommend health products, such as protein and healthy food.

MapMyRun [12] is a GPS running and workout tracking software application under iOS. It can track the fitness distance and count the calorie burnt during walking and jogging [12]. The fitness stats of users can be shared over social networking platform, such as Facebook, Twitter, Wechat, etc. MapMyRun can also rank the fitness stats among social friends to provide incentive for user's future fitness plan and schedule. Users can also share their fitness experiences for healthcare and training.

GPSports Inc. [13] offers a wide range of sports live monitoring systems for the players of football, hockey, rugby, etc. The sports live monitoring system adopts the dedicated vests for players to measure the player's live distance and speeds, which reflect their performance, such as running distance and area in football game. It also measures real-time physiology parameters, such as heart rate to maximize the athletic performance, and body load to analyze volume, intensity and work rate. The corresponding analysis system provides the comprehensive analysis of players and helps the coach to adjust the team strategy. The historical data can be used for training and guideline of future strategy.

1.3 Aim of the Monograph

Although MHN applications provide numerous opportunities and benefits, they raise various security and privacy issues. Since the health information, e.g., phenomena, health condition, emergency, is relatively sensitive for users, any inappropriate disclosure may violate user privacy and even result in the property loss [3]. Users may also worry about their critical health data to be tampered when their health data are stored in the untrusted cloud servers [14]. Moreover, some malicious attackers misbehave in MHNs to disrupt the effectiveness or mislead other users' preferences [15]. Without appropriate security and privacy protections, users may not accept MHN applications.

In addition, the costs of security protections vary with users' diverse demands, and may impact users' experiences in MHN applications. For example, complicated encryption techniques may offer users more security guarantees but with higher computational overheads and latency than lightweight ones. To satisfy users' diverse security requirements and balance the trade off between the performance and security protections, Quality-of-Protection (QoP) becomes a newly emerging security concept that allows applications to seamlessly integrate the adjustable security protections [16, 17]. Therefore, a special spotlight covering MHNs trends in security and privacy protections from QoP perspective, which can separate security schemes into different levels to ensure the suitable security services for the best trade off between performance demands and security.

In this monograph, we investigate security and privacy issues in MHNs. We introduce the overall architecture of MHN and present some promising MHN applications. Then, we discuss the security and privacy challenges in MHNs, including privacy leakage, misbehaviors, and security in health data collection and processing. We also present some solutions, i.e., privacy-preserving health data aggregation, misbehavior detection, and secure health data processing to address these security and privacy challenges. From QoP perspective, the security countermeasures for MHNs can also be adjusted to satisfy users' diverse requirements about service, experience, and protection. Finally, we present some open problems and indicate future research directions.

1.4 Organization of the Monograph

The remainder of the monograph is organized as follows. Chapter 2 discusses security and privacy challenges in MHNs, including general security and privacy requirements and some unique issues in MHNs. To address these issues, Chap. 3 proposes a priority based health data aggregation scheme with privacy preservation for cloud assisted WBANs to reduce the aggregation overheads and preserve user privacy during health data collection. Chapter 4 presents misbehavior detection scheme for health data sharing among MHN users to differentiate Sybil attackers from normal users. Chapter 5 introduces the promising health data processing techniques and presents the corresponding privacy preservation schemes in MHNs. Chapter 6 investigates attribute based access control for MHNs. Finally, Chap. 7 summarizes the monograph and proposes some future research directions in MHNs.

References

1. K. Zhang, K. Yang, X. Liang, Z. Su, X. Shen, and H. Luo, "Security and privacy for mobile healthcare networks — from quality-of-protection perspective," *IEEE Wireless Communications*, vol. 22, no. 4, pp. 104–112, 2015.
2. Forbes. [Online]. Available: http://www.forbes.com/
3. X. Liang, M. Barua, L. Chen, R. Lu, X. Shen, X. Li, and H. Luo, "Enabling pervasive healthcare through continuous remote health monitoring," *IEEE Wireless Communications*, vol. 19, no. 6, pp. 10–18, 2012.
4. K. Kiyono, Z. Struzik, N. Aoyagi, and Y. Yamamoto, "Multiscale probability density function analysis: non-gaussian and scale-invariant fluctuations of healthy human heart rate," *Biomedical Engineering, IEEE Transactions on*, vol. 53, no. 1, pp. 95–102, 2006.
5. M. Baumert, V. Baier, S. Truebner, A. Schirdewan, and A. Voss, "Short- and long-term joint symbolic dynamics of heart rate and blood pressure in dilated cardiomyopathy," *Biomedical Engineering, IEEE Transactions on*, vol. 52, no. 12, pp. 2112–2115, 2005.
6. JuniperResearch. [Online]. Available: http://www.juniperresearch.com/press/press-releases/smart-wearables-market-to-generate-$53bn-hardware
7. Q. Shen, X. Liang, X. Shen, X. Lin, and H. Luo, "Exploiting Geo-Distributed Clouds for a E-Health Monitoring System With Minimum Service Delay and Privacy Preservation," *IEEE Journal of Biomedical and Health Informatics*, vol. 18, no. 2, pp. 430–439, 2014.
8. R. Lu, X. Lin, and X. Shen, "SPOC: A secure and privacy-preserving opportunistic computing framework for mobile-healthcare emergency," *IEEE Transactions on Parallel and Distributed Systems*, vol. 24, no. 3, pp. 614–624, 2013.
9. A. Toninelli, R. Montanari, and A. Corradi, "Enabling secure service discovery in mobile healthcare enterprise networks," *IEEE Wireless Communications*, vol. 16, no. 3, pp. 24–32, 2009.

10. R. Lu, X. Lin, X. Liang, and X. Shen, "A Secure Handshake Scheme with Symptoms-Matching for mHealthcare Social Network," *MONET*, vol. 16, no. 6, pp. 683–694, 2011.
11. G. Cardone, A. Corradi, L. Foschini, and R. Montanari, "Socio-technical awareness to support recommendation and efficient delivery of IMS-enabled mobile services," *IEEE Communications Magazine*, vol. 50, no. 6, pp. 82–90, 2012.
12. MapMyRun. [Online]. Available: https://itunes.apple.com/ca/app/map-my-run-gps-running-workout/id291890420?mt=8
13. GPSports. [Online]. Available: http://gpsports.com/
14. J. Zhou, Z. Cao, X. Dong, X. Lin, and A. Vasilakos, "Securing m-healthcare social networks: challenges, countermeasures and future directions," *IEEE Wireless Communications*, vol. 20, no. 4, pp. 12–21, 2013.
15. H. Wang, D. Peng, W. Wang, H. Sharif, H. Chen, and A. Khoynezhad, "Resource-aware secure ECG healthcare monitoring through body sensor networks," *IEEE Wireless Communications*, vol. 17, no. 1, pp. 12–19, 2010.
16. C. Ong, K. Nahrstedt, and W. Yuan, "Quality of protection for mobile multimedia applications," in *Proc. of IEEE ICME*, 2003, pp. 137–140.
17. A. Luo, C. Lin, K. Wang, L. Lei, and C. Liu, "Quality of protection analysis and performance modeling in IP multimedia subsystem," *Computer Communications*, vol. 32, no. 11, pp. 1336–1345, 2009.

Chapter 2
Security and Privacy Challenges in MHN

The objective of the monograph is to investigate security and privacy in MHNs. This chapter first present the general security and privacy requirements in MHNs. Then, we discuss several unique security and privacy challenges for MHNs.

2.1 Security and Privacy Requirements

In MHNs, some general security and privacy requirements should be satisfied as follows.

1. *Integrity* should be guaranteed such that the health data measured, transmitted, shared, stored and processed in MHNs are true representations of the intended information. These data should not be tampered in any way.
2. *Confidentiality* should be ensured such that the health data from patients, hospital, service providers and some other entities are kept undisclosed and invisible to the unauthorized entities. Only the authorized entities in MHNs can reach the raw data from users.
3. *Authentication* should be achieved such that any involved entity requesting access in MHNs is authentic and valid. In MHNs, not only the information provided by the users and healthcare service providers but also the identities of the entities using such information should be verified. Any invalid information and identities should be efficiently detected.
4. *Privacy* refers to personally identifiable information or other sensitive information which should be protected for each legitimate entity in MHNs. The transmitted, stored and processed information of this legitimate entity in MHNs should be prevented from disclosing to active and passive attackers.
5. *Non-repudiation* resists the repudiation threats where attackers deny after performing some certain activities in MHNs. For example, a doctor who diagnoses

© Springer International Publishing Switzerland 2015
K. Zhang, X. Shen, *Security and Privacy for Mobile
Healthcare Networks*, Wireless Networks, DOI 10.1007/978-3-319-24717-5_2

the patient's disease may deny the diagnosis results; the examiner may deny the examine of a certain patient. MHNs should be able to detect these repudiation threats.

6. *Access Control* is to enforce access policies and ensure that only authorized users can have access to resources in MHNs. As users' private health data are stored in the cloud sever, they should be able to define access policy.
7. *Anonymity* guarantees that any involved entity cannot be identified. For instance, identities of the patients can be made anonymous when they store their health data on the cloud so that the cloud servers could not learn about the identity.
8. *Unlinkability* refers to the use of resources or items of interest multiple times by a user without other users or subjects being able to interlink the usage of these resources. More specifically, the information obtained from different flows of the health data should not be sufficient to establish linkability by the unauthorized entities [1].
9. *Accountability* is an obligation to be responsible in light of the agreed upon expectations. The patients or the entities nominated by the patients should monitor the use of their health information whenever that is accessed at hospitals, pharmacies, insurance companies etc.
10. *Audit* ensures that all the healthcare data are secure and all the data access activities in the e-Health cloud are being monitored [2].

2.2 Security and Privacy Challenging Issues in MHNs

With the main driver of user experience [3] and security service requirements, QoP becomes a prestigious security concept to provide various security protection levels for different levels of users with diverse requirements. In specific, MHNs with QoP can achieve access privileges via authentication; guarantees confidentiality, integrity, non-repudiation via encryption and signature; ensures the copyright via watermarking; protects user's privacy via cryptography, anonymity and obfuscation techniques as shown in Fig. 2.1 [3]. Having a set of security protection services, QoP can adjust these tunable protections according to different requirements since QoP is fueled by artifacts, human intelligence and involvements. Besides these off-the-shelf security protection schemes applied in QoP, several other emerging approaches should also be developed from QoP perspective to address these critical security and privacy issues in MHNs.

2.2.1 Privacy Leakage

Privacy leakage is a critical issue in MHNs as the sensitive health data are involved in the collection, transmission, processing and sharing. Without appropriate privacy protections, users may not be willing to expose their data visible to others, which

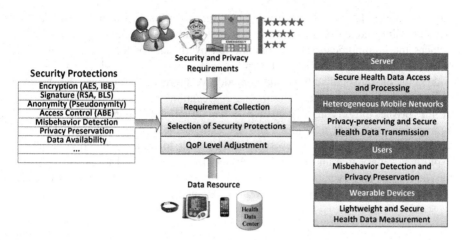

Fig. 2.1 QoP in MHNs

hinders the processing and sharing of health data and users' experiences. For example, a user suffering HIV/AIDS do not want other people to know his disease when using MHNs. The inappropriate or unconscious information leakage may leave negative impacts on this user. Another example is about fitness and training application. As mentioned in Chap. 1, football players wear dedicated vests which measure their performance for coaches to adjust the team strategy. If the measured information is leaked to the opposite team, the opposite can change strategy in advance. Therefore, privacy should be preserved in MHNs to provide user-friendly services.

In [4], several general privacy threats in healthcare system, such as identity privacy, information leakage during transmission and location privacy, are investigated. In [5], the privacy protection is applied between sensors and smartphones to protect sensing data disclosure. In [6], Ong et al. investigate the security services partitioned into various security levels to balance the security requirements and performance preferences. A proper QoP construction can be offered by the characterization of QoP with security settings, where it expresses security constraints and attributes to customize protections for different applications. In MHNs, to achieve a higher privacy level of data and users' profiles (or attributes), e.g., personal physiology parameters, the privacy protections should be robust and strong enough to resist the potential attacks and leakage, which inevitably increases the computational overheads and latency. Therefore, QoP should be applied into MHNs for adjusting the privacy protections with the data and user privacy levels.

During the health data transmission, privacy preservation is of paramount importance in MHNs. Privacy-preserving aggregation schemes are also widely investigated in recent years. Shi et al. [7] introduce a privacy-preserving aggregation of time series data which slices the data to mix them together and confines the aggregator's decryption capability, where it enables the aggregator to only decrypt the sum of the data without learning any exact data value. Lu et al. [8] utilize the increasing sequence to mix the user's multi-dimensional data together which reduces the communication and computation overheads for the aggregation. Shi et al. [9] also present a privacy-preserving aggregation scheme which supports a wide range of statistical additive and non-additive aggregation functions. Further, it can resist the collusion attack during the aggregation. To improve the robustness of the privacy-preserving aggregation, Chan et al. [10] upgrade the existing aggregation with fault tolerance. The TA assigns the aggregator N capabilities corresponding to the N low level users. The fault tolerant aggregation scheme explores a binary tree and establishes groups for the low level users to improve the robustness. However, the overheads during the user-aggregator communications are not negligible when multi-hop forwarding is involved in cloud assisted WBANs.

2.2.2 Health Data Access Control

As MHNs take the advantages of the powerful storage and computation capabilities from the outsourced cloud servers, security concerns associated with these untrusted cloud servers are also raised in MHNs. The health data access policy should be clearly defined and used to authenticate the user's identity with access authority. For example, for the patient's daily health data (e.g., Electrocardiography (ECG)) stored in the cloud server, only the doctors in the neurology subject can access these data and the corresponding analysis results. Meanwhile, the data should be protected from being accessed by insurance company [11]. Besides the general access control policies, it is still critical to ensure the fine-grained access in accordance to users' attributes. In [12], Yu et al. propose fine-grained access control for cloud storage to prevent users' sensitive data from disclosure to other untrusted servers and unauthorized users. This fine-grained access control scheme is based on ABE technique which associates the data access policy with attributes. It also reduces users' overheads by delegating the majority of computations to the powerful cloud servers.

In MHNs, the dynamic access management is necessary to address the issues of users' attribute changing, revocation, new user's participation, etc. Delegation of access control is another important issue of access control in MHNs. For example, the family doctor has the access to his patient's health records. When this family doctor needs to share the patient's health records to an expert for further diagnosis, the access control should achieve both delegation and resist the collusion attack. Least but not last, the computation overheads of access control schemes should be considered for different MHN applications. With an increasing number of attributes (which are used in attribute-based access control system), the encryption and decryption overheads may increase correspondingly. According to different access levels, the computation burden of users should be released.

2.2.3 Privacy-Preserving Health Data Processing

When the health data are outsourced to the cloud servers for further processing and analysis, users' raw data should be invisible to the untrusted cloud servers and other unauthorized entities. Furthermore, users' (e.g., data owners') identities and their personal profiles should also be anonymous to other entities according to users' privacy requirements.

Intuitively, the private information from MHN users can be encrypted such that only ciphertexts of this information are stored in the cloud servers. However, the encrypted health data hinder the processing and analysis at the cloud server's side. To address this issue, some secure multi-party computation schemes (e.g., functional encryption, homomorphic encryption) are proposed to guarantee the data protection during some basic operations, e.g., aggregation, summation and comparison. With different QoP requirements, the protections should be enhanced when applying some complicated operations, such as Bayesian learning, data mining, which are essential for health data analysis and diagnosis. In addition, when applying homomorphic encryption to machine learning or data mining algorithms, it consumes a large amount of computation and communication overheads, which may considerably reduce the battery lifetime of wearable devices and smartphones. The increased computation and communication overheads can also increase the delay of health data analysis. Therefore, it is challenging to balance the trade off between health data analysis and privacy preservation for machine learning and data mining algorithms.

Recently, Bost et al. [13] develop a set of secure machine learning classification schemes and propose a library of components. Yuan et al. [14] propose a collaborative learning scheme, which enables each user to encrypt his data and upload the ciphertext to the cloud server. The cloud server performs most of the learning algorithms over these ciphertext without learning the plaintext. A variant of "doubly homomorphic" encryption scheme for secure multi-party computation is adopted to perform flexible operations over the encrypted data.

2.2.4 Malicious Attacks and Misbehaviors

MHN, as a new paradigm expanding the traditional Internet to a ubiquitous network connecting wearable devices in the physical world, starts an evolution to enhance the interaction among people [15–17]. With the emerging wireless communication techniques, such as short range wireless communications and WiFi, MHNs enable users to share information with others in social network applications [18–20]. Furthermore, by integrating the sensing, communication and computation capabilities [21], MHNs can offer diverse intelligent services [22] to form smart home [23], smart community [24] and smart city [25, 26] as shown in Fig. 2.2. With the advancement of MHNs, these value-added applications flourish to facilitate people to interact with others and change the way we communicate with each other.

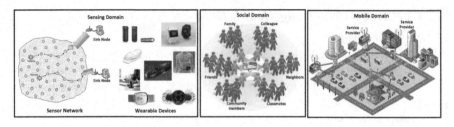

Fig. 2.2 MHN domains: sensing domain, social domain, and mobile domain

However, the emerging MHN is vulnerable to Sybil attacks where attackers can manipulate fake identities [27–29] or abuse pseudo identities to compromise the effectiveness of the systems. In the presence of Sybil attacks, MHNs might generate incorrect medical reports, while users may receive spam and leak their private information. According to a statistical report [30] in 2012, a substantial number of user accounts are detected as fake (or Sybil) accounts in online social networks (OSNs). About 76 million (7.2 %) accounts are Sybil in Facebook, while 20 million newly-created fake accounts join Twitter per week. These Sybil accounts may broadcast advertisements and spam, or even disseminate fishing websites and malware in OSNs to steal other users' private information. In a mobile social networking systems [31], Sybil attackers produce various biased options with "legible" accounts. Without an effective detection in mobile networks, the collective results would be easily manipulated by the attackers. Since most of Sybil attackers have similar behaviors as normal users, it is still difficult to detect them.

According to different types of Sybil attacks in MHNs, we define them in three types. At the beginning, we present the social graph model. Suppose an undirect social graph denoted as $\mathcal{G}$ with n honest nodes H and totally m edges. Sybil nodes are denoted as S. In the social graph, we use node to represent user, identity, or account in the real network. The edge between every pair of two nodes is weighted by their social relations. An attack edge AG is the edge connecting an honest node and a Sybil one, as shown in Fig. 2.3. Note that in some literatures [32, 33], social network refers to the undirect social graph $\mathcal{G}$.

- *SA-1 Sybil Attacks*:

The SA-1 attackers usually build connections within the Sybil community as shown in Fig. 2.3, i.e., Sybil nodes tightly connect with other Sybil nodes. However, the SA-1's capability of building social connections with honest nodes is not strong. In other words, the number of social connections between Sybil nodes and honest ones is limited, i.e., in Fig. 2.3, the number of SA-1 attack edges is limited.

The SA-1 attackers usually exist in sensing domain and social domain, i.e., OSN, voting [34] or mobile sensing systems [35]. The main goal is to manipulate the overall option or popularity. For example, in an online voting system, SA-1 can illegally forge a massive number of identities to perform as normal users and submit the votes with the biased options. The final voting result might be manipulated by the SA-1 attackers, since a considerable portion of votes are from the SA-1 attackers.

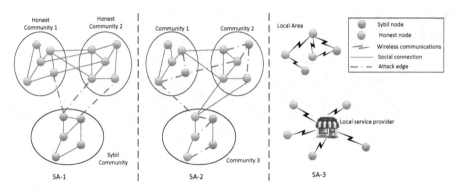

Fig. 2.3   Three types of Sybil attacks: SA-1, SA-2 and SA-3

Similarly, in mobile sensing system, SA-1 can forge the false sensing data and indirectly change the aggregated data. Therefore, in some cases, the behaviors of Sybil attackers are indistinguishable from the normal users.

- *SA-2 Sybil Attacks*:

SA-2 attackers usually exist in social domain. Different from SA-1, SA-2 is able to build the social connections not only among Sybil identities but also with the normal users. In other words, the capability of SA-2 is strong to mimic the normal user's social structures from the perspective of social graph. Therefore, the number of attack edges is large.

The goal of SA-2 is to disseminate spam, advertisements, and malware; steal and violate user's privacy; and maliciously manipulate the reputation system. For example, in OSNs, SA-2 can forge the profiles and friend list as normal users, but purposely spread spam, advertisements and malware. In addition, SA-2 could generate plenty of positive review comments in a service evaluation system to exaggerate the advantages of service, or generate many negative comments to underestimate services. Obviously, SA-2 would focus on some specific behaviors and repeat them in the high frequency [36].

- *SA-3 Sybil Attacks*:

There are SA-3 Sybil attackers in mobile networks (i.e., mobile domain). The primary goal of SA-3 is similar to that of SA-2. However, the impact of SA-3 may be in a local area or within a short period. Due to the dynamics of mobile networks [37, 38], mobile users cannot keep connections with others for the long time, or the connections are intermittent. Furthermore, the centralized authority cannot exist in mobile networks at all the time. Thus, unlike that in the online system, the social relationships, global social structure, topology and historical behavior patterns in mobile networks are not easy to obtain for Sybil defense towards SA-3. The lack of the global information and dynamic mobility of mobile users leads to the difficulties in detecting SA-3 compared with the detection of SA-1 and SA-2. We compare different types of Sybil attacks in Table 2.1.

Table 2.1 Three types of Sybil attacks

Categories of Sybil attacks	Social graph features	Attack goal	Behavior discrimination	Mobility
SA-1	Sybils exist in the same region or community, and the number of attack edges is limited	Maliciously or purposely upload the biased reports or comments (positive or negative) to manipulate the overall option and dominate the whole system	Perform as the normal users, and repeat specific behaviors frequently	×
SA-2	Sybils may tightly connect with normal users, and generate more attack edges	Disseminate spam and malware to launch some other attacks, camouflage as normal users, or violate other users' privacy	Purposely repeat some specific behaviors in the high frequency	×
SA-3	Sybils may tightly connect with normal users	Manipulate the local popularity, disseminate spam in the mobile environment, or violate user's privacy	Repeat specific behaviors frequently	√

According to the aforementioned discussion, Sybil attackers can misbehave in different patterns and act as similarly as normal users, posing new challenges to Sybil detection. Therefore, an effective Sybil detection scheme is essential for MHNs to guarantee users' security and privacy requirements.

2.3 Summary

In this chapter, we have discussed security and privacy challenges in MHNs. First, we have introduced general security and privacy requirements for MHNs, such as integrity, confidentiality, authentication, privacy, non-repudiation, access control, anonymity, unlinkability, accountability and audit. Then, we have discussed several security and privacy challenges in MHNs, including privacy leakage, health data access control, privacy-preserving data processing and misbehaviors. In the following chapters, we will present some countermeasures to overcome these challenging issues in MHNs.

References

1. C. Lai, H. Li, X. Liang, R. Lu, K. Zhang, and X. Shen, "CPAL: A Conditional Privacy-Preserving Authentication With Access Linkability for Roaming Service," *IEEE Internet of Things Journal*, vol. 1, no. 1, pp. 46–57, 2014.
2. A. Abbas and S. U. Khan, "A Review on the State-of-the-Art Privacy-Preserving Approaches in the e-Health Clouds," *IEEE Journal of Biomedical and Health Informatics*, vol. 18, no. 4, pp. 1431–1441, 2014.
3. M. Katsarakis, G. Fortetsanakis, P. Charonyktakis, A. Kostopoulos, and M. Papadopouli, "On user-centric tools for QoE-based recommendation and real-time analysis of large-scale markets," *IEEE Communications Magazine*, vol. 52, no. 9, pp. 37–43, 2014.
4. J. Zhou, Z. Cao, X. Dong, X. Lin, and A. Vasilakos, "Securing m-healthcare social networks: challenges, countermeasures and future directions," *IEEE Wireless Communications*, vol. 20, no. 4, pp. 12–21, 2013.
5. H. Wang, D. Peng, W. Wang, H. Sharif, H. Chen, and A. Khoynezhad, "Resource-aware secure ECG healthcare monitoring through body sensor networks," *IEEE Wireless Communications*, vol. 17, no. 1, pp. 12–19, 2010.
6. C. Ong, K. Nahrstedt, and W. Yuan, "Quality of protection for mobile multimedia applications," in *Proc. of IEEE ICME*, 2003, pp. 137–140.
7. E. Shi, T. Chan, E. Rieffel, R. Chow, and D. Song, "Privacy-preserving aggregation of time-series data," in *Proc. of NDSS*, 2011, pp. 1–17.
8. R. Lu, X. Liang, X. Li, X. Lin, and X. Shen, "EPPA: An efficient and privacy-preserving aggregation scheme for secure smart grid communications," *IEEE Transactions on Parallel and Distributed Systems*, vol. 23, no. 9, pp. 1621–1631, 2012.
9. J. Shi, R. Zhang, Y. Liu, and Y. Zhang, "PriSense: Privacy-preserving data aggregation in people-centric urban sensing systems," in *Proc. IEEE INFOCOM*, 2010, pp. 758–766.
10. T. Chan, E. Shi, and D. Song, "Privacy-preserving stream aggregation with fault tolerance," *IACR Cryptology ePrint Archive*, vol. 2011, p. 655, 2011.
11. M. Barni, P. Failla, R. Lazzeretti, A. Sadeghi, and T. Schneider, "Privacy-preserving ECG classification with branching programs and neural networks," *IEEE Transactions on Information Forensics and Security*, vol. 6, no. 2, pp. 452–468, 2011.
12. S. Yu, C. Wang, K. Ren, and W. Lou, "Achieving Secure, Scalable, and Fine-grained Data Access Control in Cloud Computing," in *Proc. of IEEE INFOCOM*, 2010, pp. 534–542.
13. R. Bost, R. Popa, S. Tu, and S. Goldwasser, "Machine Learning Classification over Encrypted Data," in *Proc. of NDSS*, 2015, pp. 1–14.
14. J. Yuan and S. Yu, "Privacy Preserving Back-Propagation Neural Network Learning Made Practical with Cloud Computing," *IEEE Transactions on Parallel and Distributed Systems*, vol. 25, no. 1, pp. 212–221, 2014.
15. A. Sehgal, V. Perelman, S. Kuryla, and J. Schonwalder, "Management of resource constrained devices in the Internet of Things," *IEEE Communications Magazine*, vol. 50, no. 12, pp. 144–149, 2012.
16. K. Ren, W. Lou, K. Zeng, and P. Moran, "On broadcast authentication in wireless sensor networks," *IEEE Transactions on Wireless Communications*, vol. 6, no. 11, pp. 4136–4144, 2007.
17. Y. Liu, K. Liu, and M. Li, "Passive diagnosis for wireless sensor networks," *IEEE/ACM Transactions on Networking*, vol. 18, no. 4, pp. 1132–1144, 2010.
18. K. Zhang, X. Liang, X. Shen, and R. Lu, "Exploiting multimedia services in mobile social networks from security and privacy perspectives," *IEEE Communications Magazine*, vol. 52, no. 3, pp. 58–65, 2014.
19. L. Foschini, T. Taleb, A. Corradi, and D. Bottazzi, "M2M-based metropolitan platform for IMS-enabled road traffic management in IoT," *IEEE Communications Magazine*, vol. 49, no. 11, pp. 50–57, 2011.

20. W. He, G. Yan, and L. Xu, "Developing vehicular data cloud services in the IoT environment," *IEEE Transactions on Industrial Informatics*, to appear.
21. C. Perera, A. Zaslavsky, P. Christen, and D. Georgakopoulos, "Context aware computing for the Internet of Things: A survey," *IEEE Communications Surveys Tutorials*, vol. 16, no. 1, pp. 414–454, 2014.
22. H. Celdran, G. Clemente, G. Perez, and M. Perez, "SeCoMan: A semantic-aware policy framework for developing privacy-preserving and context-aware smart applications," *IEEE Systems Journal*, to appear.
23. J. Huang, Y. Meng, X. Gong, Y. Liu, and Q. Duan, "A novel deployment scheme for green Internet of Things," *IEEE Internet of Things Journal*, to appear.
24. X. Li, R. Lu, X. Liang, X. Shen, J. Chen, and X. Lin, "Smart community: an Internet of Things application," *IEEE Communications Magazine*, vol. 49, no. 11, pp. 68–75, 2011.
25. J. Jin, J. Gubbi, S. Marusic, and M. Palaniswami, "An information framework of creating a smart city through Internet of Things," *IEEE Internet of Things Journal*, to appear.
26. P. Vlacheas, R. Giaffreda, V. Stavroulaki, D. Kelaidonis, V. Foteinos, G. Poulios, P. Demestichas, A. Somov, A. Biswas, and K. Moessner, "Enabling smart cities through a cognitive management framework for the Internet of Things," *IEEE Communications Magazine*, vol. 51, no. 6, pp. 102–111, 2013.
27. Q. Lian, Z. Zhang, M. Yang, Y. Zhao, Y. Dai, and X. Li, "An empirical study of collusion behavior in the Maze P2P file-sharing system," in *Proc. of IEEE ICDCS*, 2007, pp. 56–66.
28. M. Yang, Z. Zhang, X. Li, and Y. Dai, "An empirical study of free-riding behavior in the Maze P2P file-sharing system," in *Proc. of IPTPS*, 2005, pp. 182–192.
29. K. Zhang, X. Liang, R. Lu, and X. Shen, "Exploiting multimedia services in mobile social network from security and privacy perspectives," *IEEE Communications Magazine*, vol. 52, no. 3, pp. 58–65, 2014.
30. Businessinsider. [Online]. Available: http://www.businessinsider.com/
31. K. Zhang, X. Liang, R. Lu, K. Yang, and X. Shen, "Exploiting mobile social behaviors for sybil detection," in *Proc. of IEEE INFOCOM*, 2015, pp. 271–279.
32. H. Yu, M. Kaminsky, P. Gibbons, and A. Flaxman, "SybilGuard: Defending against sybil attacks via social networks," *IEEE ACM Transactions on Networking*, vol. 16, no. 3, pp. 576–589, 2008.
33. H. Yu, P. Gibbons, M. Kaminsky, and F. Xiao, "SybilLimit: A near-optimal social network defense against sybil attacks," *IEEE/ACM Transactions on Networking*, vol. 18, no. 3, pp. 885–898, 2010.
34. D. Tran, B. Min, J. Li, and L. Subramanian, "Sybil-Resilient online content voting," in *Proc. of NSDI*, 2009, pp. 15–28.
35. Y. Reddy, "A game theory approach to detect malicious nodes in wireless sensor networks," in *Proc. of SENSORCOMM*, 2009, pp. 462–468.
36. G. Wang, T. Konolige, C. Wilson, X. Wang, H. Zheng, and B. Zhao, "You are How You Click: Clickstream analysis for sybil detection," in *Proc. of USENIX*, 2013, pp. 241–255.
37. X. Liang, K. Zhang, R. Lu, X. Lin, and X. Shen, "EPS: An Efficient and Privacy-Preserving Service Searching Scheme for Smart Community," *IEEE Sensors Journal*, vol. 13, no. 10, pp. 3702–3710, 2013.
38. Q. Xu, Z. Su, K. Zhang, P. Ren, and X. Shen, "Epidemic Information Dissemination in Mobile Social Networks with Opportunistic Links," *IEEE Transactions on Emerging Topics in Computing*, vol. 3, no. 3, pp. 399–409, 2015.

Chapter 3
Secure Health Data Collection in MHN

In this chapter, we propose a priority based health data aggregation scheme (PHDA) with privacy preservation for cloud assisted WBANs to reduce the aggregation overheads and preserve user privacy. The health data are divided into different types, and each type of data is assigned a specific priority. When a user wants to upload his data, he can select different forwarding strategies according to his data's priority. The intuition is that the data with higher priority can be forwarded in a smaller delay. Furthermore, the data with the same priority can be efficiently aggregated which significantly reduces the communication overheads.

3.1 Introduction

In MHNs, Wireless Body Area Networks (WBANs) plays an essential role in real time monitoring. As the phenomenon of aging population and the demands of remote health monitoring in our daily life [1], users wear the portable sensors and wearable devices which form a WBAN to measure patients and users' physiology parameters, such as temperature, blood pressure, ECG, etc. These measured data are first transmitted to smartphones or other terminals, which forward them to the hospital. WBANs provide a wide range of healthcare services, such as health condition monitoring, sports and fitness, entertainments, and military applications [2]. With the ever-increasing demands from patients and users, it is necessary to timely process the sensed data and obtain diagnosis (or feedback) from doctors. Since the health data volume keeps increasing, it requires many network resources, such as bandwidth, storage, computation and communication power, which poses challenges for the traditional WBANs to achieve the above goals [3]. Thus, cloud computing [4] becomes a new paradigm to assist WBANs to store and process the sensed health data.

© Springer International Publishing Switzerland 2015
K. Zhang, X. Shen, *Security and Privacy for Mobile
Healthcare Networks*, Wireless Networks, DOI 10.1007/978-3-319-24717-5_3

As the cloud server can store a large number of sensed health data and process them for guideline to doctor's diagnosis [5, 6], cloud assisted WBANs offer the desired services for patients and achieve reliability of MHNs. For example, in a conference environment, sport stadium and social community, people have certain social activities [7] and wear WBANs [8] to sense their real-time health data which are periodically reported to the cloud server. The doctors in hospital, as trusted authorities, can access his patient's health data via the cloud server. Under such a circumstance, doctors are able to timely diagnose the abnormal symptom and return the diagnosis results to patients. If a user has an emergency, WBANs can automatically call the hospital and continuously upload his real-time physiology parameter to help doctors judge the patient's health condition. However, when a large number of users staying in the same area simultaneously upload their health data to the cloud server, the connections between cloud servers and WBANs may be intermittent. The available bandwidths of each user are also limited such that the MHN performance is significantly degraded. Therefore, one communication bottleneck of MHN locates between WBANs and cloud servers and poses challenges to achieve reliability and efficiency.

From the literatures [9, 10], users' cooperation is adopted to improve the communication reliability. Liang et al. [11] propose an emergency call scheme for healthcare applications and epidemically spread health data to enable these data to be fast delivered to the cloud server and hospitals. Although this scheme can improve the delivery ratio and reduce delay of emergency calls, the communication overheads are considerably high. From the aforementioned example, physiology parameters of the patients with the emergency should be continuously measured and forwarded to the cloud server for the monitoring and diagnosis. If this portion of data are still epidemically spread in MHNs, numerous network resources are consumed such that MHNs may be even disabled. To address this issue, the health data should be classified into various types with corresponding communication strategies based on different requirements, such as delay and delivery ratio.

During the communications of cloud assisted WBAN, security concerns are raised [12]. All the transmitted data in MHNs should be authenticated and with security protections against malicious deletion and modifications. For instance, a malicious attacker may forge an emergency call and let it spread in MHNs to block other users' communications. Meanwhile, privacy is also a primary concern from user's perspective, since health data are usually highly sensitive. For example, the ECG signal reflects people's certain behaviors, such as having meals and sleeping. The reveal of ECG signal may help attackers to identify the user's daily behaviors and violate user's privacy [13]. Therefore, it is still challenging to efficiently collect health data and preserve user's privacy in cloud assisted WBANs.

In specific, the main contributions of this chapter are three-fold as follows.

- Firstly, we propose a priority based data aggregation scheme in cloud assisted WBANs. The health data are classified into different types assigned with corresponding priorities. Various health data forwarding strategies are selected based on the health data priority. Furthermore, social spots are used to facilitate the PHDA and help users forward health data to the cloud server. An eligible relay is selected based on his social tie to the social spots, reflecting the relay's forwarding capability.

- Secondly, we propose a lightweight privacy-preserving aggregation scheme with aggregate authentication. The cloud server can only learn the statistical information among the aggregated users without knowing the exact data from individual user. Furthermore, the proposed aggregate authentication scheme can verify users' health data priority and resist the forgery attack. It also reduces the authentication overhead.
- Finally, we discuss the security and privacy features of the PHDA and show that it cannot only preserve identity privacy and data privacy but also resist the forgery attack. Meanwhile, the performance evaluation shows that the PHDA can meet the requirements of delivery ratio and delay for health data with different priorities. It also consumes lower communication overheads compared with other schemes.

The remainder of this chapter is organized as follows: The related works are reviewed in Sect. 3.2. Network model and design goals are presented in Sect. 3.3. In Sect. 3.4, we propose the PHDA scheme, followed by the security analysis and performance evaluation in Sects. 3.5 and 3.6, respectively. Finally, Sect. 3.7 summarizes the chapter.

3.2 Related Work

Recently, there are several research works [14–16] on data forwarding in mobile ad hoc networks. Some of them investigate and exploit social features, such as community [17], centrality [18], betweenness [19], and social proximity [20] to make use of the intermittent connections among mobile users for efficient data forwarding. Exploiting centrality and community, Hui et al. [18] propose a social based routing protocol to purposely forward the packets to the users with some social relationship with the destination. With the consideration of time-varying social features, Gao et al. [21] study the transient contact distribution, transient connectivity and transient community structure to develop an efficient data forwarding protocol with these novel metrics. They observe that the users' behaviors vary from time to time, and users might belong to different social communities at different time. Wu et al. [22] propose a social based multi-path routing in MSNs, which utilizes entropy to analyze user's social features for efficient routing. Bulut et al. [23] adopt the concept of friendship to measure the degree of users' direct and indirect contact, which can improve the forwarding efficiency. However, security and privacy issues should also be addressed for data forwarding. Most of forwarding schemes require a large amount of social or contact information such that some private and sensitive information is disclosed. Costantino et al. [24] investigate the trade off between privacy and forwarding accuracy, where the privacy leakage is formulated as the entropy.

Another type of effective forwarding scheme is to deploy several fixed nodes in advance to help mobile users forward their data. Aviv et al. [14] investigate the user's mobility patterns and propose a social spot based forwarding scheme, i.e., "Return-to-Home", which allows the fixed social spots to help mobile users store-and-forward the packets and improves the forwarding efficiency. Lu et al. [10] propose a social spot aided packet forwarding scheme (SPRING) with privacy preservation in VANETs. The SPRING follows the "Return-to-Home" principle and preserves user privacy at the same time. Zhang et al. [16] investigate a novel social spot deployment strategy to preserve the location privacy for users and social spots. Despite the social spots, our PHDA also enables mobile users to help forward the data to social spots so that the data forwarding efficiency is further improved. In addition, some research efforts are paid to investigate data forwarding in healthcare systems. Borrego et al. [25] investigate a new paradigm, called store-carry-process-and-forward, based on mobile code to improve the integration of wireless sensor networks and grid computing infrastructures. Liang et al. [11] propose a privacy-preserving emergency call scheme, named PEC, for mobile healthcare social networks. The PEC exploits the epidemic dissemination for emergency call and provides fine-grained access control on the emergency data. In terms of aggregation, Yager [26] proposes the priority based data aggregation scheme with multiple criteria aggregation. In [26], the trade-off between the data priority and satisfaction to criteria is investigated. Yager also proposes two schemes to formulate the priority based aggregation with multiple criteria. Misra et al. [27] consider the bandwidth shifting and redistribution problems for mobile cloud with the QoS guarantee. They utilize the gateway to aggregate the demands from the mobile users, and formulate it as an utility maximization problem. Misra et al. [28] propose a lightweight energy-efficient routing scheme for wireless sensor network to increase the network life time. The neighbor nodes with higher energy aggregates the data from other nodes and forwards the aggregated data packet to the destination.

Privacy-preserving aggregation schemes are also widely investigated in recent years. Shi et al. [29] introduce a privacy-preserving aggregation of time series data which slices the data to mix them together and confines the aggregator's decryption capability, where it enables the aggregator to only decrypt the sum of the data without learning any exact data value. Lu et al. [30] utilize the increasing sequence to mix the user's multi-dimensional data together which reduces the communication and computation overheads for the aggregation. Shi et al. [31] also present a privacy-preserving aggregation scheme which supports a wide range of statistical additive and non-additive aggregation functions. Further, it can resist the collusion attack during the aggregation. To improve the robustness of the privacy-preserving aggregation, Chan et al. [32] upgrade the existing aggregation with fault tolerance. The TA assigns the aggregator N capabilities corresponding to the N low level users. The fault tolerant aggregation scheme explores a binary tree and establishes groups for the low level users to improve the robustness. However, the overheads during the user-aggregator communications are not negligible when multi-hop forwarding is involved in cloud assisted WBANs. Therefore, the priority based privacy-preserving aggregation scheme is very important in terms of efficiency.

3.3 Problem Definition

In this section, we first describe the network model and identify our primary design goals to establish a reliable and secure connection between WBANs and cloud servers. Then, we present the attacker model and illustrate the security requirements.

3.3.1 Network Model

We consider a cloud assisted WBAN consisting of a trusted authority, L social spots, a small portion of semi-trusted cloud servers, and N mobile patients/users as shown in Fig. 3.1. The details of these entities are presented as follows.

- *Trusted Authority (TA)* is a trustable, powerful, and storage-rich entity, and bootstraps the whole system in the initialization phase. In the real world, the TA could be a certificated hospital having the responsibility to manage the users' health data. When bootstrapping, the TA generates secret keys for each legitimate user, and users' certificates for further authentication. After the aggregation, the TA can decrypt the data from each individual user for diagnosis. Upon receiving attack reports from residential users, the TA revokes the malicious users and adjusts the users' encryption keys.
- *Social Spot (SS)* is a pre-deployed local gateway and equipped with storage-rich and powerful communication devices. According to the user's behaviors, totally L social spots are located at the intersections or spots where a large portion of mobile users visit frequently. The SS directly collects the health sensing data

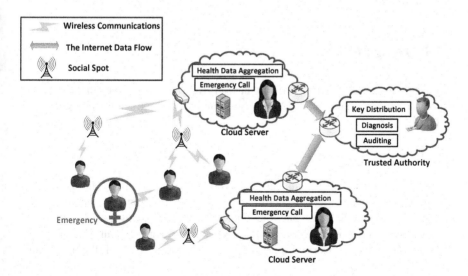

Fig. 3.1 Network model for cloud assisted WBAN

from each individual user via WBAN communications. Finally, the *SS*s upload the aggregated data to the cloud servers via the Internet.

- *Cloud Server (CS)* stores the large volume of health sensing data from mobile users, and processes some data, such as ECG, to produce the useful information for doctor's diagnosis. Since some third parties, e.g., insurance company, can access the CS for query and some other operations, the cloud server is a semi-trusted entity in cloud assisted WBAN. To achieve data confidentiality and user's privacy, the data stored in the CSs are of ciphertexts.
- *Mobile users* are denoted by $\mathbb{U} = \{u_1, u_2, \ldots, u_N\}$. Each mobile user is equipped with body area sensors which monitors the personal health sensing data in a real time fashion and periodically uploads these health data to the CS via the user's smartphone or PDA [33]. At the beginning, the individual user u_i should firstly register to the TA for the profiles (unique identity), certificates and key materials. Then, u_i should keep them secure and generate session keys in each time slot. When u_i obtains sensing data or faces an emergency, u_i only needs to forward the corresponding data to any one of *SS*s.

3.3.2 Attacker Model

Malicious users might exist in the network and launch attacks to violate legitimate user's identity and data privacy, and degrade the network performance. Some inside users might forge the data priority, i.e., making a fake emergency call, or increasing their data priority, to degrade the network performance. Furthermore, the cloud server is semi-trusted, and some third parties might launch attacks on the cloud servers to violate user's data privacy.

3.3.3 Design Goals

Our design goal is to develop a priority based privacy-preserving health data aggregation scheme for cloud assisted WBANs to improve the aggregation efficiency.

3.3.3.1 Efficiency Goals

We intend to reduce the communication overhead of the aggregation, and guarantee the delivery ratio and delay according for the data in different priorities. The health sensing data should be classified into different types with specific priorities. For different data priorities, the data forwarding strategies should be different and maximize the network resource usage with satisfaction of the minimum requirements.

3.3.3.2 Security Goals

Our primary security goal is to protect the individual user's data from disclosure and resist the forgery attack.

- **Data Privacy**: The proposed scheme should not only refine the cloud server's decryption capability [34] but also protect the user's data from eavesdropping during the communications. Therefore, the individual user's data privacy should be protected.
- **Identity Privacy**: The legitimate users might not want to disclose their unique identity information, especially when they are close to the social spots. Therefore, the proposed scheme should be able to prevent the malicious users or attackers from identifying them.
- **Forgery Attack**: A malicious user could forge a false emergency call, or increase his data priority so that his data can be preferentially uploaded to the cloud server. The proposed scheme should be able to detect the forged data and block them in the network.

3.4 Proposed PHDA Protocol

In this section, we first provide an overview of the PHDA. Then, we present the details of our proposed PHDA scheme, which mainly consists of initialization, health data generation, and priority based data aggregation.

3.4.1 Overview of PHDA

To efficiently aggregate user's health sensing data, we investigate the priority based data forwarding in cloud assisted WBANs. Towards a variety of sensing data, different forwarding strategies should be selected to not only forward data within the given delay but also consumes the reasonable network resources.

First of all, we classify the health data into three categories: emergency call, vital health data, and regular health data. As depicted in Table 3.1, the emergency call is the highest priority data and should be successfully delivered to the cloud server as fast as possible. In addition, the time line is divided into many small time periods. At the beginning of each period, every user obtains his/her health data from the wearble WBANs. The vital health data are the requested data by doctors for continuous monitoring on the user with the emergency. The regular data are not for the emergency user so that the delay requirement is not that tough. Usually, they should be delivered to the cloud server before the next time period. We further divide the vital and regular data into small data and big data. The small data, such as physiology parameters with the size of 10–100 bytes, should be delivered to the

Table 3.1 Data priority in cloud assisted WBANs

Priority	Data category	Data size
P_5	Emergency call	Small
P_4	Vital physiology parameter	Small
P_3	Vital image data	Large
P_2	Regular physiology parameter	Small
P_1	Regular image data	Large

Table 3.2 Frequently used notations

Notation	Definition
B_i	Buffer size of user u_i
$d_{i,j}$	Data of priority j from user u_i
$\mathbb{EM}_i$	Emergency on user u_i
$|P_{j,i}|$	Data size of priority j from user u_i
ST_{u_i}	Social tie of u_i
TH_v	Threshold of forwarding P_4 and P_3 data
TH_r	Threshold of forwarding P_3 and P_2 data
TH_E	Threshold of forwarding emergency call

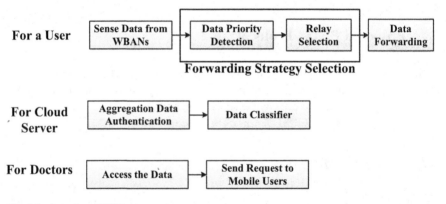

Fig. 3.2 Overview of PHDA

cloud server within a given delay. On the other hand, the big data, such as ECG or images, are of large size, and should be uploaded in time without consuming too many network resources. Some frequently used notions are depicted in Table 3.2.

A mobile user u_i sets his data priority with the data priority detection module as shown in Fig. 3.2. According to the data priority, u_i has different forwarding strategies to forward his data. With our proposed relay selection algorithm, the optimal relay can be selected for different data priorities. When the mobile users visits any one of the pre-deployed social spots, the data can be forwarded to SSs and finally uploaded to the cloud servers.

For the cloud server, the CS first authenticates and classifies the aggregated data. Then, the data can be accessed by different entities, including hospital, doctors, insurance companies. The doctors can send request to some specific mobile users via the CSs for vital data monitoring. With the request from the doctors, the mobile users can make their vital data verified when forwarding them to other relays.

Specifically, the PHDA can be proceeded in the following phases: initialization phase, health data generation, priority based data aggregation, and data decryption.

3.4.2 Initialization

The TA, such as the authorized hospital or healthcare center, initializes the network and audits the aggregated data. We utilize bilinear pairing [35] and Paillier cryptograph [36] to achieve the privacy-preserving aggregation.

Let $\mathbb{G}, \mathbb{G}_1$ be two additive cyclic groups of the same prime order q, and P be the generator of $\mathbb{G}$. There is a bilinear map $e : \mathbb{G} \times \mathbb{G} \longrightarrow \mathbb{G}_1$. A bilinear pairing exists if it is computationally efficient for $e(aP_1, bP_2) = e(P_1, P_2)^{ab} \in \mathbb{G}_1$ for any $P_1, P_2 \in \mathbb{G}$ and all $a, b \in \mathbb{Z}_q^*$, and $e(P, P) \neq 1_{\mathbb{G}_1}$. A bilinear key generation algorithm $Gen(\kappa)$ is used to produce the key materials, where κ is the system security parameter of bilinear pairing.

During the system initialization, the TA first generates $(q, P, \mathbb{G}, \mathbb{G}_1, e)$ by running the key generation algorithm $Gen(\kappa)$. Then, the TA selects the Paillier cryptographic security parameter κ' and two large primes p', q' where $|p'| = |q'| = \kappa'$. The public keys of Paillier cryptograph are: (1) $n = p' \cdot q'$; (2) $g \in \mathbb{Z}_{n^2}^*$ as the generator. The secret keys are: (1) $\lambda = \mathsf{lcm}(p' - 1, q' - 1)$ where lcm the least common multiple of $p' - 1$ and $q' - 1$; (2) $\mu = \frac{1}{\mathsf{L}(g^\lambda \bmod n^2)} \bmod n$ where L is a defined function and $\mathsf{L}(x) = \frac{x-1}{n}$.

Suppose the maximum number of health data with each priority from N users is smaller than a constant ϕ. The data value for each priority is less than a constant θ. Then, the TA builds up a superincreasing sequence [30] $\overrightarrow{b} = (b_1 = 1, b_2, \cdots, b_N)$, where b_i is a large prime, the length $|b_i| \geqslant \kappa$. $\sum_{j=1}^{i-1} b_j \cdot \phi \cdot \theta < b_i$ for $i = 2, \cdots, N$, and $\sum_{j=1}^{N} b_j \cdot \phi \cdot \theta < n$. Similarly, the TA builds up another superincreasing sequence $\overrightarrow{a} = (a_1 = 1, a_2, \cdots, a_5)$, where $a_2, \cdots, a_5$ are large primes and the length $|a_i| \geqslant \kappa$. Let $\sum_{i=1}^{N} b_i = \gamma$. We have $\sum_{j=1}^{i-1} a_j \cdot \gamma \cdot \theta < a_i$ for $i = 2, \cdots, 5$, and $\sum_{j=1}^{5} a_j \cdot \gamma \cdot \theta < n$. Finally, the TA obtains $(g_1, g_2, \cdots, g_5)$ where $g_i = g^{a_i}$ for $i = 1, 2, \cdots, 5$.

Afterwards, the TA selects a random number $x \in Z_q^*$ to compute $\mathsf{Y} = xP$ as the public key. $\mathsf{H} : \{0, 1\}^* \longrightarrow \mathbb{G}$ and $\mathsf{H}_1 : \{0, 1\}^* \longrightarrow Z_q^*$ are cryptographic hash functions. The secret keys are $(\lambda, \mu, \overrightarrow{a}, \overrightarrow{b}, \alpha, x)$. The public keys are $\{q, P, \mathbb{G}, \mathbb{G}_1, e, n, g_1, g_2, g_3, g_4, g_5, \mathsf{Y}, \mathsf{H}, \mathsf{H}_1\}$.

When a user u_i registers to the TA, u_i obtains his secret keys b_i. With the multiple pseudonym techniques [37], u_i is also assigned with a set of asymmetric key pairs and uses the alternatively changing public keys as the user's pseudonyms PID_i for the communications. The unique identity u_i can be protected as only literally-meaningless pseudonyms are exposed to the public. u_i selects a random number $x_i \in \mathbb{Z}_q^*$ as his private key (Fig. 3.3).

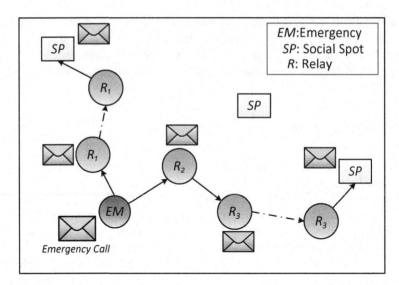

Fig. 3.3 Forwarding of emergency call

3.4.3 Health Data Generation

WBANs worn on or in the user's body can measure the physiology parameters and some large-size sensing data (i.e., ECG). The user u_i should forward the data with a specific priority to the cloud server within a given deadline which is the maximum delay for the specific data priority. Here, the data $(d_1, d_2, d_3, d_4, d_5)$ are generated with different priorities. u_i first chooses a random number $r_i \in Z_q^*$ and computes

$$C_{i,j} = g_j^{b_i d_{i,j}} \cdot r_i^n \bmod n^2, \text{ where } j \in \{1, 2, 3, 4, 5\}. \tag{3.1}$$

Here, $j \in \{1, 2, 3, 4, 5\}$ is the priority number. Note that the ciphertexts for different data priorities cannot be combined together because the forwarding strategies are different. But the ciphertext from the same data priority can be combined together. u_i then signs on the data with his private key x_i to generate the signature. For the regular data, u_i records the system time *Time* and makes the signature $\mathsf{R}_{i,j}$

$$\mathsf{R}_{i,j} = x_i \mathsf{H}(C_{i,j} || \mathsf{PID}_i || \textit{Time}) \bmod n^2 \tag{3.2}$$

Regarding the vital data, the TA chooses a random number $s \in \mathbb{Z}_q^*$, and computes $S = sP$. Then, the TA sends $\mathsf{REQ}_i || S$ to user u_i where $\mathsf{REQ}_i = \mathsf{H}_1(\text{Data Type} || \textit{Time})$. With REQ_i, u_i can authenticate his data priority as the vital level (P3 or P4). u_i makes the signature $\mathsf{V}_{i,j}$ on the vital data $d_{i,j}$ with his private key x_i as

$$\mathsf{V}_{i,j} = x_i S + x_i \mathsf{H}_1(C_{i,j}) \mathsf{Y}. \tag{3.3}$$

3.4.4 Priority Based Data Aggregation

After the data generation, a user u_i wants to forward his data as soon as possible. From the view of the network, we have to balance the traffic and optimize the network recourses. Therefore, we propose a priority based data aggregation scheme to not only guarantee the forwarding delay but also reduce the communication overheads. We provide the different forwarding strategies for the data with different priorities.

(1) Emergency Call: When a user u_i has an emergency event denoted as $\mathbb{EM}_i = (u_i\|Des_i\|Time\|Location)$, where Des_i is the general description of the emergency, u_i sets his data priority as P5 in Table 3.1. When u_i meets another user u_r, u_i first checks whether the social tie ST_{u_r} is larger than u_i's or the difference between ST_{u_r} and ST_{u_i} is less than the threshold of emergency call TH_E. If one of the conditions holds, u_r is selected as an emergency relay. Then, u_i makes short group signature [38, 39] G.sign(u_i) and forwards $(\mathbb{EM}_i\|$G.sign$(u_i))$ to u_r.

Receiving $(\mathbb{EM}_i\|$G.sign$(u_i))$, u_r first checks the signature G.sign(u_i) with G.verify(G.sign(u_i)). Here, we use G.sign and G.verify to denote the group signature and verification algorithms. If invalid, u_r reports u_i to the SS or TA. If valid, u_r carries and forwards $\mathbb{EM}_i$ to any SS. Once u_r meets another user u_{r_1} before u_r forwards $\mathbb{EM}_i$ to SS, u_r follows the steps that u_i does and determines whether $\mathbb{EM}_i$ should be forwarded to u_{r_1} or not.

When u_r visits any SS, u_r forwards $\mathbb{EM}_i$ to SS. Since all SSs are connected via the Internet, the data can be successfully uploaded if one SS receives the data.

(2) Vital Data Forwarding: When u_i is encountered with u_r and has a piece of vital data $C_{i,j}$ where $j = 3$ or 4, u_i checks whether u_r meets the following criteria: (1) the social tie ST_{u_r} is larger than u_i's or the difference between ST_{u_r} and ST_{u_i} is less than the threshold (TH_v) of P4 or P3; and (2) the available buffer size of u_r is larger than the transmitting data size. If both conditions hold, u_r is selected as a relay. Then, u_i forwards data to u_r.

After receiving the data, u_r first checks the signature. If invalid, u_r reports u_i to the SS or TA. If valid, u_r computes $C_{r,j} = C_{r,j}C_{i,j} \bmod n^2$, and forwards $C_{r,j}$ to any SS if possible. Once u_r meets another user u_{r_1} before u_r forwards the data to SS, u_r follows the steps that u_i does and determines whether the data should be forwarded to u_{r_1} or not.

(3) Regular Data Forwarding: For P1 and P2 data, u_i also needs to check whether a relay u_r is eligible or not by following: (1) the available buffer size of u_r is larger than u_i's data; (2) the social tie ST_{u_r} of u_r is larger than ST_{u_i}, or $ST_{u_i} - ST_{u_r} \leqslant TH_r$. If and only if both conditions hold, u_i can forward the data to u_r. The detailed relay selection steps are depicted in Algorithm 1.

Algorithm 1 Relay selection

1: Two users u_s and u_r are encountered, and u_s has an emergency.
2: **if** $ST_{u_r} > ST_{u_s}$ OR $ST_{u_s} - ST_{u_r} > TH_E$ **then**
3:   u_s forwards the emergency call $\mathbb{EM}_s$ to u_r AND u_r verifies the emergency call $\mathbb{EM}_s$.
4:   **if** $\mathbb{EM}_s$ is valid **then**
5:    u_r stores $\mathbb{EM}_s$ and forwards it to AP or another relay if possible AND $B_r = B_r - |\mathbb{EM}_s|$.
6:    **if** u_s has P$_4$ data AND $B_r > 0$ **then**
7:     **if** $ST_{u_r} > ST_{u_i}$ OR $ST_{u_i} - ST_{u_r} < TH_v$ **then**
8:      u_s forwards its P$_4$ data to P$_{2,s}$ u_r AND $B_r = B_r - |P_{4,s}|$.
9:     **end if**
10:    **end if**
11:    **if** u_s has P$_3$ data AND $B_r > |P_{3,s}|$ **then**
12:     **if** $ST_{u_r} > ST_{u_i}$ OR $ST_{u_i} - ST_{u_r} < TH_v$ **then**
13:      u_s forwards its P$_3$ data P$_{3,s}$ to u_r AND $B_r = B_r - |P_{3,s}|$.
14:     **end if**
15:    **end if**
16:    **if** u_s has P$_2$ data AND $B_r > 0$ **then**
17:     **if** $ST_{u_r} > ST_{u_i}$ OR $ST_{u_i} - ST_{u_r} < TH_r$ **then**
18:      u_s forwards its P$_2$ data to u_r AND $B_r = B_r - |P_{2,s}|$.
19:     **end if**
20:    **end if**
21:    **if** u_s has P$_1$ data AND $B_r > |P_{5,s}|$ **then**
22:     **if** $ST_{u_r} > ST_{u_i}$ OR $ST_{u_i} - ST_{u_r} < TH_r$ **then**
23:      u_s forwards its P$_1$ data to u_r AND $B_r = B_r - |P_{5,s}|$.
24:     **end if**
25:    **end if**
26:   **else**
27:    u_r reports u_s as a malicious user to the TA.
28:   **end if**
29: **end if**
30: **End Procedure**

3.4.5 Data Aggregation for the Cloud Server

(1) Aggregate Authentication: When the *CS* receives $M \leqslant N$ vital data in time period t, the *CS* first verifies the authenticity of the data. First, the *CS* computes the sum of all the signatures $\sum_{i=1}^{N} V_i$, and checks

$$e\left(\sum_{i=1}^{M} V_i, P\right) \stackrel{?}{=} e(S, \sum_{i=1}^{M} PK_i) \cdot \prod_{i=1}^{M} e(Y, PK_i)^{H_1(C_{i,j})}. \tag{3.4}$$

If Eq. 3.4 holds, all M vital data packets are authenticated. The correctness can be proved as

$$e\left(\sum_{i=1}^{M} V_{i,j}, P\right) = \prod_{i=1}^{M} e(x_i S + x_i H_1(C_{i,j}) x P, P)$$

$$= \prod_{i=1}^{M} e(S, x_i P) \cdot \prod_{i=1}^{M} e(x H_1(C_{i,j}) P, x_i P) \tag{3.5}$$

$$= e(S, \sum_{i=1}^{M} \mathsf{PK}_i) \cdot \prod_{i=1}^{M} e(x H_1(C_{i,j}) P, \mathsf{PK}_i)$$

$$= e(S, \sum_{i=1}^{M} \mathsf{PK}_i) \cdot \prod_{i=1}^{M} e(Y, \mathsf{PK}_i)^{H_1(C_{i,j})}$$

Here, S is distributed by the TA in the time period t, PK_i is u_i's public key, and Y is the TA's public key. The pairing operations $e(S, \sum_{i=1}^{M} \mathsf{PK}_i)$ and $e(Y, \mathsf{PK}_i)$ can be pre-computed. During each aggregate authentication, only one pairing operation is required so that the authentication efficiency is considerably improved.

For the regular data, the TA can do the batch verification to efficiently verify the signatures as

$$e\left(\sum_{i=1}^{N} R_{i,j}, P\right) = e(\sum_{i=1}^{N} x_i H(C_{i,j}||\mathsf{PID}_i||Time), P) \tag{3.6}$$

$$= \prod_{i=1}^{N} e(H(C_{i,j}||\mathsf{PID}_i||Time), \mathsf{PK}_i)$$

When the CS receives all the data from N mobile users at the end of every time slot (one time period is divided into several time slots), the CS aggregates all the ciphertexts $C_{i,j}$ together, and sends $C = \prod_{j=1}^{5} \prod_{i=1}^{N} C_{i,j} \bmod n^2$ to the TA. In addition, the CS generates the signature of the aggregated data C as

$$\mathsf{Sign}_{CS} = x_{CS} H(C||CS||Time) \bmod n^2 \tag{3.7}$$

where x_{CS} is the private key of the CS.

3.4.6 Data Decryption by the TA

After receiving the aggregated data from the CS, the TA first verifies the signature Sign_{CS}. The TA checks whether

$$e(\mathsf{Sign}_{CS}, P) \stackrel{?}{=} e(H(C||CS||Time), \mathsf{PK}_{CS}). \tag{3.8}$$

If it holds, the received data are valid.

The TA has the data as

$$C = \prod_{j=1}^{5} g_j^{\sum_{i=1}^{N} b_i d_{ij}} \left(\prod_{i=1}^{N} r_i \right)^{n} \mod n^2$$

$$= g_1^{\sum_{i=1}^{N} b_i d_{i1}} \cdot g_2^{\sum_{i=1}^{N} b_i d_{i2}} \cdots g_5^{\sum_{i=1}^{N} b_i d_{i5}} \cdot \left(\prod_{i=1}^{N} r_i \right)^{5n} \mod n^2$$

$$= g^{a_1 \sum_{i=1}^{N} b_i d_{i1}} \cdot g^{a_2 \sum_{i=1}^{N} b_i d_{i2}} \cdots g^{a_5 \sum_{i=1}^{N} b_i d_{i5}} \cdot \left(\prod_{i=1}^{N} r_i \right)^{5n} \mod n^2 \qquad (3.9)$$

$$= g^{a_1 \sum_{i=1}^{N} b_i d_{i1} + a_2 \sum_{i=1}^{N} b_i d_{i2} + \cdots + a_5 \sum_{i=1}^{N} b_i d_{i5}} \cdot \left(\prod_{i=1}^{N} r_i \right)^{5n} \mod n^2$$

Let $M = a_1 \sum_{i=1}^{N} b_i d_{i1} + a_2 \sum_{i=1}^{N} b_i d_{i2} + \cdots + a_5 \sum_{i=1}^{N} b_i d_{i5}$, and $r = \left(\prod_{i=1}^{N} r_i \right)^{5}$, we have $C = g^M r^n \mod n^2$. The TA can use his secret key (λ, μ) to decrypt the aggregated data according to Paillier cryptograph.

Having the aggregated data, the TA runs Algorithm 2 to decrypt the data for each priority. The correctness of Algorithm 2 can be achieved as

$$X_l = a_1 \sum_{i=1}^{N} b_i d_{i1} + a_2 \sum_{i=1}^{N} b_i d_{i2} + \cdots + a_{l-1} \sum_{i=1}^{N} b_i d_{il}$$

$$< a_1 \sum_{i=1}^{N} \theta + a_2 \sum_{i=1}^{N} \theta + \cdots + a_{l-1} \sum_{i=1}^{N} \theta \qquad (3.10)$$

$$= \sum_{j=1}^{l-1} a_j N \theta < a_l$$

Algorithm 2 Recover the aggregated data

1: Input: $\vec{a} = (a_1 = 1, a_2, \cdots, a_N)$ and M
2: Output: $D_1, D_2, \cdots, D_l$
3: Set $X_l = M$
4: **for** $j = l$ to 2 **do**
5:    $X_{j-1} = X_l \mod a_j$
6:    $D_j = \frac{X_j - X_{j-1}}{a_j} = \sum_{i=1}^{N} b_i d_{ij}$
7: **end for**
8: $D_1 = X_1 = \sum_{i=1}^{N} b_i d_{i1}$
9: Return $(D_1, D_2, \cdots, D_l)$
10: **End Procedure**

Therefore, we have $X_{l-1} = X_l \mod a_l = a_1 \sum_{i=1}^{N} b_i d_{i1} + a_2 \sum_{i=1}^{N} b_i d_{i2} + \cdots + a_{l-1} \sum_{i=1}^{N} b_i d_{il-1}$ and

$$\frac{X_l - X_{l-1}}{a_l} = \frac{a_l \sum_{i=1}^{N} d_{il}}{a_l} = \sum_{i=1}^{N} d_{il} = D_l \,, \text{ where } l = 1, 2, \cdots, 5. \tag{3.11}$$

X_l is the sum for the data with the l-th priority. Similarly, $d_{i,j}$ is obtained by the TA.

3.5 Security Analysis

In this section, we discuss the security properties of our proposed PHDA scheme. We focus on the aforementioned security requirements in Sect. 3.3.

- **Data Privacy**: The user's data privacy can be achieved based on the assumption that Decisional Diffie-Hellman (DDH) problem or Decisional Bilinear Diffie-Hellman (DBDH) [40] problem is hard. The passive eavesdropping can be resisted since all the transmitted data are encrypted by Paillier cryptograph. Furthermore, the whole superincreasing sequences $\overrightarrow{a}$ and $\overrightarrow{b}$ are the secret keys which are securely kept by the TA. Each user u_i can only obtain b_i for the encryption. As a result, the cloud server and other entities including the attackers who do not know all the secret keys cannot recover the exact data for different priorities. Therefore, the user's data privacy can be achieved. Due to the properties of the superincreasing sequence, the TA can recover the data even though they are aggregated together.
- **Identity Privacy**: The user's identity privacy can be preserved with the multiple pseudonym techniques. In each time period, a user u_i changes his pseudonym PID_i to protect his identity privacy. Only the meaningless pseudonyms are exposed to the other users. No entity except the TA can trace the pseudonyms of u_i and link them together. By frequently changing his pseudonyms, u_i can protect his identity privacy due to the unlinkability of the current pseudonym and the previous ones. On the other hand, if u_i launches some attacks, the TA is able to trace u_i's pseudonyms PID_i and link them together to identify the attacker.
- **Forgery Attack**: The malicious insider users cannot launch forgery attack to tamper with the data priority because desired vital sensing data are transmitted with a request $\mathsf{REQ}_i \| S$ from the TA. All the collected vital data should be verified by the TA and authenticate that the data type is exactly the one the doctors require. If the malicious user U forges a signature

$$\mathsf{V}_\mathsf{U} = x_\mathsf{U} S' + x_\mathsf{U} \mathsf{H}_1(C_\mathsf{U}) \mathsf{Y}. \tag{3.12}$$

The other users can verify it and have

$$e(V_U, P) \neq e(S, \mathsf{PK_U}) \cdot e(Y, \mathsf{PK_U})^{\mathsf{H}_1(C_U)}. \tag{3.13}$$

Then, U is revoked by the TA and drawn to the revocation list.

In addition, the emergency call cannot be forged by the outside attackers due to the unforgeability of the group signature is adopted. Only the registered user can obtain the key materials from the TA to produce the valid emergency call signature. If an attacker A forges an emergency call $\mathbb{EM}_A$, other legitimate users can verify A's signature with $\mathsf{G.verify(G.sign}_(A))$ and detect the attack.

In summary, from the above discussions, the PHDA can resist the forgery attack from the inside malicious users and the outside attackers.

3.6   Performance Evaluation

3.6.1   Computational Complexity

We compare the computational complexity of the PHDA with the non-aggregate scheme. In the PHDA, an individual user u_i encrypts the health data with 6 exponentiation operations in $\mathbb{Z}_{n^2}$. For the signature generation, u_i performs 1 and 2 multiplication operations $\mathbb{G}$ for regular health data and vital data, respectively. The cloud server CS needs to verifies the signatures of the received health data. The vital data verification requires $M + 2$ paring operations, which are the primary computational costs, and M exponentiation operations in $\mathbb{G}_1$, and M multiplication operations in $\mathbb{G}_1$. Meanwhile, CS performs $N + 1$ pairing operations and N multiplication operations in $\mathbb{G}_1$. When sending the aggregated data to the TA, the CS generates the signature with 1 multiplication operation in $\mathbb{G}$. The multiplication operations in $\mathbb{Z}_{n^2}$ can be considered negligible compared with the exponentiation, paring operations. Therefore, the overhead of data aggregation can be negligible. TA verifies the signature with 2 paring operations, and decrypts the data from CS with 1 exponentiation operations in $\mathbb{Z}_{n^2}$.

We compare the proposed PHDA scheme with the non-aggregate scheme where the data are directly sent to the TA in the separate type. The individual user u_i needs to separately encrypt 5 types of health data with 10 exponentiation operations in $\mathbb{Z}_{n^2}$, and generate signatures with 5 multiplication operations in $\mathbb{G}_1$. At the TA end, it requires $10N$ paring operations for verification and $5N$ exponentiation operations in $\mathbb{Z}_{n^2}$ to decrypt the data.

The computational complexity of PHDA and non-aggregate scheme is depicted in Table 3.3. We denote C_e as the exponentiation operation in $\mathbb{Z}_{n^2}$, C_m as the multiplication operation in $\mathbb{G}$, $C_{m,1}$ as the multiplication operation in $\mathbb{G}_1$, C_p as the paring operation. As depicted in Table 3.3, the computation overhead of the TA is significantly reduced with the assistance of the cloud server.

Table 3.3 The comparison of computational complexity between PHDA and non-aggregate scheme

	PHDA	Non-aggregate scheme
TA	$2C_p + C_e$	$10NC_p + 5NC_e$
Cloud server	$(M+N+3)C_p + (M+N)C_{m,1} + C_m$	N/A
Users	$6C_e + 3C_m$	$10C_e + 5C_m$

3.6.2 Simulation Setup

For the simulation, we utilize a real world human trace—Infocom06 [41] trace, where 78 mobile users attend a conference within 4 days. Every two mobile users' encounter in the proximity can be detected via their attached Bluetooth devices. There are several fixed nodes in the trace, and we use them as the social spots according to their contacts with mobile users. Finally, we select 10 fixed nodes as social spots in our simulation. The contacts of all users and fixed nodes are recorded in the log file. For the simulation, we collect 128,979 useful contacts, and divide them into two portions: the first one third of the data set as a training set producing user's social ties and the residual data as the experiment set used for the simulation. We implement the PHDA and some other schemes under the Matlab simulator to evaluate the performance. Basically, we utilize delivery ratio, average delay and number of copies as metrics for the comparison.

3.6.3 Simulation Results

To evaluate the emergency call's forwarding efficiency of the PHDA, we implement the PHDA, Epidemic and SPRING schemes for comparison. The Epidemic forwarding, which enables every encountered user to forward the data, is also adopted in some other emergency call schemes [11]. The SPRING [10] only relies on mobile users to forward their own data to the social spots. Totally 78 emergency calls are generated randomly. The comparison results shown in Figs. 3.4, 3.5 and 3.6 with the comparison among PHDA, Epidemic and SPRING schemes in terms of delivery ratio, average delay and the number of copies. From Fig. 3.4, the delivery ratio of the PHDA is less than that of the Epidemic at the beginning of the emergency event. However, with the PHDA, 85 % emergency calls can be successfully forwarded to the servers within 2 min, while the percentage for the Epidemic is around 90 %. The PHDA and Epidemic can achieve the same delivery ratio after 6 min and finally reach 100 % delivery. Regarding the SPRING, it consumes less communication overhead but cannot achieve the desirable delivery ratio which is not suitable for healthcare applications. From Fig. 3.6, we can see that the communication overhead of the PHDA is significantly reduced compared with the Epidemic. The reason is

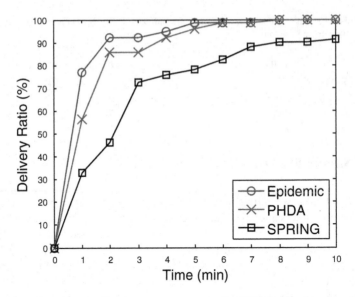

Fig. 3.4 Delivery ratio comparison between emergency call of PHDA and epidemic

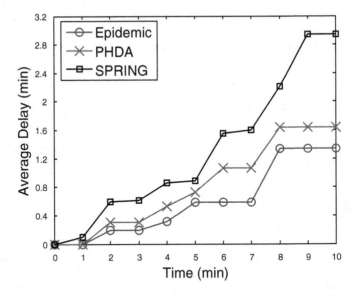

Fig. 3.5 Average delay comparison between emergency call of PHDA and epidemic

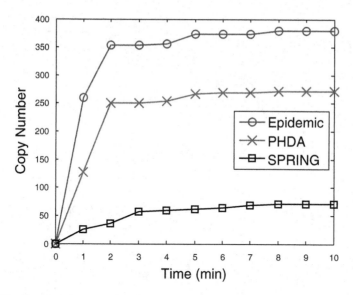

Fig. 3.6 Number of copies comparison between emergency call of PHDA and epidemic

that the PHDA utilizes the fixed social spots to help mobile users store-and-forward the data so that the fixed social spots provide more opportunities for mobile users to forward their data. Furthermore, the deployment of the social spots is selected at the location where a lot of mobile users visit frequently. In addition, the PHDA enables the mobile users to select the active mobile users which further improve the connections between the mobile users and social spots. Therefore, the delivery ratio of the PHDA is close to the Epidemic with much lower communication overhead.

In Figs. 3.7, 3.8 and 3.9, we show the impact of the copy constraints on the PHDA with a constant social tie constraint *TH*. Here, the copy constraint is the maximum number of copies that a user can hold. With this constraint, any mobile user cannot take too many copies which significantly save each individual user's storage and energy consumption. Therefore, the network resources are fairly utilized. With a lower copy constraint, for example, at most 3 packets can be held by a user, the delivery ratio is less than that with a higher copy constraint from Fig. 3.7. But after copy constraint reaches 7, the delivery ratio varies a little because the number of eligible relay is bounded by the social tie constraint. On the other hand, with a lower available buffer size (the maximum number of copies), the communication overhead is considerably reduced.

The impacts of the social tie threshold *TH* on the performance of the PHDA are shown in Figs. 3.10, 3.11 and 3.12. We set the copy constraint as 5. From Figs. 3.10 and 3.11, with a larger *TH*, the PHDA achieves better performance in terms of delivery ratio and average delay. But the improvement is not that high. The number of copies increases when *TH* is larger from Fig. 3.12. This is because the larger *TH* causes the increased number of eligible relays which correspondingly increase the number of copies.

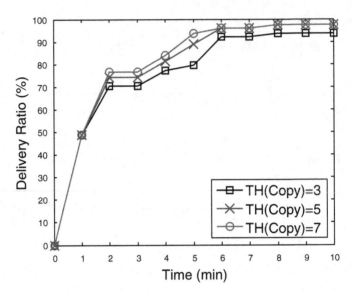

Fig. 3.7 Delivery ratio vs. copy constraint

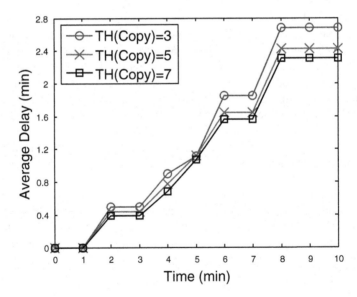

Fig. 3.8 Average delay vs. copy constraint

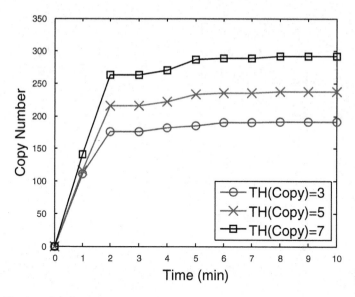

Fig. 3.9 Number of copies vs. copy constraint

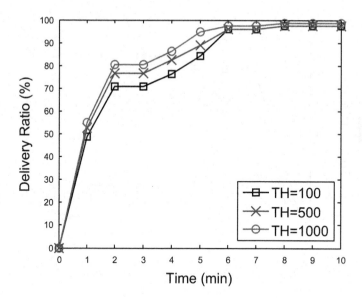

Fig. 3.10 Delivery ratio vs. TH

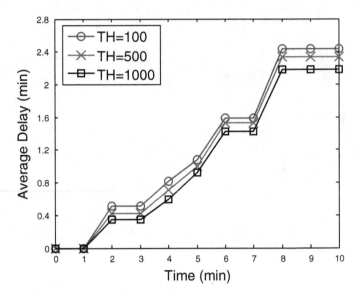

Fig. 3.11 Average delay vs. TH

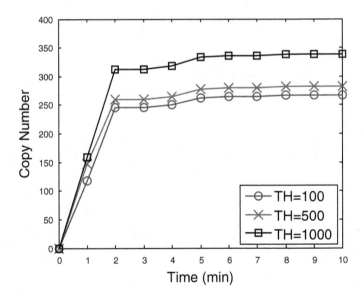

Fig. 3.12 Number of copies vs. TH

3.7 Conclusions

In this chapter, we have proposed a priority based privacy-preserving health data aggregation scheme for cloud assisted WBANs to improve the aggregation efficiency and preserve identity and data privacy. The PHDA utilizes the fixed social spots and the social tie between users and social spots to select the optimal relay and provides reliable data aggregation. With different data priorities, the forwarding strategies are adjustable and the corresponding delay requirements can be satisfied with the minimum communication overheads. The security analysis demonstrates that the PHDA can preserve identity and data privacy, while it also resists the forgery attack from inside malicious users and outside attackers. The performance evaluation shows that the PHDA satisfies the delay and delivery ratio requirements for the data with different priorities, and reduces the communication overheads at the same time. In our future work, we intend to investigate the lightweight homomorphic aggregation scheme to further reduce the communication and computation overheads.

References

1. H. Viswanathan, B. Chen, and D. Pompili, "Research challenges in computation, communication, and context awareness for ubiquitous healthcare," *IEEE Communications Magazine*, vol. 50, no. 5, pp. 92–99, 2012.
2. U. Mitra, B. Emken, S. Lee, M. Li, V. Rozgic, G. Thatte, H. Vathsangam, D. Zois, M. Annavaram, S. Narayanan, M. Levorato, D. Spruijt-Metz, and G. Sukhatme, "KNOWME: A Case Study in Wireless Body Area Sensor Network Design," *IEEE Communications Magazine*, vol. 50, no. 5, pp. 116–125, 2012.
3. J. Caldeira, J. Rodrigues, and P. Lorenz, "Toward ubiquitous mobility solutions for body sensor networks on healthcare," *IEEE Communications Magazine*, vol. 50, no. 5, pp. 108–115, 2012.
4. N. Zhang, N. Cheng, A. Gamage, K. Zhang, J. Mark, and X. Shen, "Cloud assisted hetnets toward 5g wireless networks," *IEEE Communications Magazine*, vol. 53, no. 6, pp. 59–65, 2015.
5. A. Azadeh, I. Fam, M. Khoshnoud, and M. Nikafrouz, "Design and implementation of a fuzzy expert system for performance assessment of an integrated health, safety, environment (HSE) and ergonomics system: The case of a gas refinery," *Elsevier Information Sciences*, vol. 178, no. 22, pp. 4280–4300, 2008.
6. N. Botts, B. Thoms, A. Noamani, and T. Horan, "Cloud computing architectures for the underserved: Public health cyberinfrastructures through a network of healthATMs," in *Proc. of HICSS*, 2010, pp. 1–10.
7. K. Zhang, X. Liang, R. Lu, and X. Shen, "Exploiting multimedia services in mobile social network from security and privacy perspectives," *IEEE Communications Magazine*, vol. 52, no. 3, pp. 58–65, 2014.
8. X. Liang, X. Lin, K. Zhang, and X. Shen, "Security and privacy in mobile social network: challenges and solutions," *IEEE Wireless Communications*, vol. 21, no. 1, pp. 33–41, 2014.
9. C. Liu, J. Wen, Q. Yu, B. Yang, and W. Wang, "HealthKiosk: A family-based connected healthcare system for long-term monitoring," in *Proc. of IEEE Infocom*, 2011, pp. 241–246.
10. R. Lu, X. Lin, and X. Shen, "SPRING: A social-based privacy-preserving packet forwarding protocol for vehicular delay tolerant networks," in *Proc. of IEEE INFOCOM*, 2010, pp. 632–640.

11. X. Liang, R. Lu, L. Chen, X. Lin, and X. Shen, "PEC: A privacy-preserving emergency call scheme for mobile healthcare social networks," *Journal of Communications and Networks*, vol. 13, no. 2, pp. 102–112, 2011.
12. M. Valero, S. Jung, A. Uluagac, Y. Li, and R. Beyah, "Di-Sec: A distributed security framework for heterogeneous Wireless Sensor Networks," in *Proc. of IEEE INFOCOM*, 2012, pp. 585–593.
13. K. Zhang, X. Liang, M. Barua, R. Lu, and X. Shen, "PHDA: A priority based health data aggregation with privacy preservation for cloud assisted WBANs," *Information Sciences*, vol. 284, pp. 130–141, 2014.
14. A. Aviv, M. Sherr, M. Blaze, and J. Smith, "Evading Cellular Data Monitoring with Human Movement Networks," in *USENIX Workshop on Hot Topics in Security (HotSec)*, 2010, pp. 1–6.
15. K. Zhang, X. Liang, R. Lu, and X. Shen, "Exploiting private profile matching for efficient packet forwarding in mobile social networks," *Opportunistic Mobile Social Networks*, pp. 283–312, 2014.
16. K. Zhang, X. Liang, R. Lu, X. Shen, and H. Zhao, "VSLP: Voronoi-socialspot-aided packet forwarding protocol with receiver location privacy in MSNs," in *Proc. of GLOBECOM*, 2012, pp. 348–353.
17. J. Fan, J. Chen, Y. Du, W. Gao, J. Wu, and Y. Sun, "Geo-community-based broadcasting for data dissemination in mobile social networks," *IEEE Transactions on Parallel and Distributed Systems*, vol. 24, no. 4, pp. 734–743, 2013.
18. P. Hui, J. Crowcroft, and E. Yoneki, "BUBBLE Rap: Social-based forwarding in delay-tolerant networks," *IEEE Transactions on Mobile Computing*, vol. 10, no. 11, pp. 1576–1589, 2011.
19. E. M. Daly and M. Haahr, "Social network analysis for information flow in disconnected delay-tolerant MANETs," *IEEE Transactions on Mobile Computing*, vol. 8, no. 5, pp. 606–621, 2009.
20. T. Abdelkader, K. Naik, A. Nayak, N. Goel, and V. Srivastava, "SGBR: A routing protocol for delay tolerant networks using social grouping," *IEEE Transactions on Parallel and Distributed Systems*, vol. 24, no. 12, pp. 2472–2481, 2013.
21. W. Gao, G. Cao, T. L. Porta, and J. Han, "On exploiting transient social contact patterns for data forwarding in delay-tolerant networks," *IEEE Transactions on Mobile Computing*, vol. 12, no. 1, pp. 151–165, 2013.
22. J. Wu and Y. Wang, "Social feature-based multi-path routing in delay tolerant networks," in *Proc. of IEEE INFOCOM*, 2012, pp. 1368–1376.
23. E. Bulut and B. K. Szymanski, "Exploiting friendship relations for efficient routing in mobile social networks," *IEEE Transactions on Parallel and Distributed Systems*, vol. 23, no. 12, pp. 2254–2265, 2012.
24. G. Costantino, F. Martinelli, and P. Santi, "Investigating the privacy vs. forwarding accuracy tradeoff in opportunistic interest-casting," *IEEE Transactions on Mobile Computing*, vol. 13, no. 4, pp. 824–837, 2014.
25. C. Borrego and S. Robles, "A store-carry-process-and-forward paradigm for intelligent sensor grids," *Elsevier Information Sciences*, vol. 222, pp. 113–125, 2013.
26. R. Yager, "On prioritized multiple-criteria aggregation," *IEEE Transactions on Systems, Man, and Cybernetics, Part B*, vol. 42, no. 5, pp. 1297–1305, 2012.
27. S. Misra, S. Das, M. Khatua, and M. Obaidat, "QoS-Guaranteed Bandwidth Shifting and Redistribution in Mobile Cloud Environment," *IEEE Transactions on Cloud Computing*, vol. 2, no. 2, pp. 181–193, 2014.
28. S. Misra and P. Thomasinous, "A simple, least-time, and energy-efficient routing protocol with one-level data aggregation for wireless sensor networks," *The Journal of Systems and Software*, vol. 83, no. 5, pp. 852–860, 2010.
29. E. Shi, T. Chan, E. Rieffel, R. Chow, and D. Song, "Privacy-preserving aggregation of time-series data," in *Proc. of NDSS*, 2011, pp. 1–17.
30. R. Lu, X. Liang, X. Li, X. Lin, and X. Shen, "EPPA: An efficient and privacy-preserving aggregation scheme for secure smart grid communications," *IEEE Transactions on Parallel and Distributed Systems*, vol. 23, no. 9, pp. 1621–1631, 2012.

31. J. Shi, R. Zhang, Y. Liu, and Y. Zhang, "PriSense: Privacy-preserving data aggregation in people-centric urban sensing systems," in *Proc. IEEE INFOCOM*, 2010, pp. 758–766.
32. T. Chan, E. Shi, and D. Song, "Privacy-preserving stream aggregation with fault tolerance," *IACR Cryptology ePrint Archive*, vol. 2011, p. 655, 2011.
33. "Medical Body Area Networks First Report and Order," 2009. [Online]. Available: http://www. fcc.gov/document/medical-body-area-networks-first-report-and-order
34. M. Barua, X. Liang, R. Lu, and X. Shen, "ESPAC: Enabling security and patient-centric access control for ehealth in cloud computing," *International Journal of Security and Networks*, vol. 6, no. 2/3, pp. 67–76, 2011.
35. D. Boneh and M. Franklin, "Identity Based Encryption From the Weil Pairing," in *Proc. of CRYPTO*, 2001, pp. 213–229.
36. P. Paillier, "Public-key cryptosystems based on composite degree residuosity classes," in *Proc. of EUROCRYPT*, 1999, pp. 223–238.
37. J. Freudigery, M. Manshaeiy, J. Hubauxy, and D. Parkes, "On non-cooperative location privacy: A game-theoretic analysis," in *Proc. of CCS*, 2009, pp. 324–337.
38. D. Boneh, X. Boyen, and H. Shacham, "Short group signatures," 2004. [Online]. Available: http://hovav.net/dist/groupsigs.ps
39. D. Boneh and X. Boyen, "Short signatures without random oracles and the SDH assumption in bilinear groups," *Springer-Verlag, 2008*.
40. L. Wang, L. Wang, Y. Pan, Z. Zhang, and Y. Yang, "Discrete logarithm based additively homomorphic encryption and secure data aggregation," *Elsevier Information Sciences*, vol. 181, no. 16, pp. 3308–3322, 2011.
41. J. Scott, R. Gass, J. Crowcroft, P. Hui, C. Diot, and A. Chaintreau, "CRAWDAD trace cambridge/haggle/imote/infocom (v. 2006-01-31)."

Chapter 4
Health Data Sharing with Misbehavior Detection

Despite the traditional health monitoring, MHNs can offer a wide range of social network and information sharing applications. In these social applications, MHNs are vulnerable to the malicious attacks and misbehaviors of mobile users, which may degrade the performance and even disrupt MHNs. The attackers may forge their social attributes to snatch other legitimate users' health information during the information sharing, which violates users' privacy. It may help them to push some biased health product recommendations and spam [1]. Moreover, these attackers may also misbehave, e.g., not following the network protocol, spreading spams, launching Denial-of-Service (DoS) attacks or consuming a large amount of network resources. Although some misbehavior detection schemes [2] can partially resist certain types of attacks, it is still challenging to adjust the security protection against the powerful attacks, such as Sybil attacks. The cost of misbehavior detection may increase due to the skyrocketing attacking capabilities of these attackers. To offer MHNs from QoP perspective, the misbehaviors should be categorized into different levels with the corresponding detection or protection schemes. In this chapter, we present social based mobile Sybil detection scheme to differentiate malicious attackers from normal users.

4.1 Introduction

Sybil attackers can manipulate a large number of identities to profit from services without offering sufficient contribution [3, 4]. Such misbehaviors can compromise the effectiveness of MHNs [5]. For example, Sybil attackers may spread spam, maliciously mislead the overall popularity, or violate normal user's privacy by forging a large number of fake identities. In mobile social applications, e.g., Fon11, FireChat, WeChat and Groovr, users can share and exchange their social information directly via smartphones in the local area or among the crowd.

© Springer International Publishing Switzerland 2015
K. Zhang, X. Shen, *Security and Privacy for Mobile
Healthcare Networks*, Wireless Networks, DOI 10.1007/978-3-319-24717-5_4

However, Sybil attackers can frequently change pseudonyms to repeatedly broadcast the same/similar information, such as social recommendation, traffic condition, etc. From the perspective of the encountered users, all the similar information seems to be from different senders, which may mislead the encountered user's opinions and preferences. Since mobile Sybil attackers can be merged into the crowd or rapidly move with unpredictable trajectories, it is intractable to detect them in mobile environments.

Extensive research efforts [6–8] have been put on Sybil detection by using social graph or community detection, while some works [5, 9, 10] investigate the network characteristics, such as wireless channel characteristics, or utilize cryptography to detect Sybil attackers. However, mobile users cannot easily detect Sybil attackers in mobile environments due to some limitations, e.g., the lack of strong social relationships, dynamic user mobility and limited detection capabilities. Firstly, it is difficult to establish the social graph for mobile users. Usually, mobile users may not have tight social relationships with each other in the local area. Their mobility is highly dynamic such that the social connections are hardly maintained for a long time. Without a stable social graph, some traditional social-graph based Sybil detection schemes is not applicable in mobile environments. Secondly, smarter Sybil attackers can mimic as normal users and merge into the normal user crowd or social community, which may disrupt the community-based Sybil detection. Thirdly, mobile users have limited knowledge about each other, and lack of powerful detection capabilities, such as storage and computation. To alleviate mobile user's resource consumption, one possible solution is to leverage the cloud server to assist mobile users for the data storage and computation. However, as the cloud server is generally untrusted by the mobile users, critical security and privacy concerns are raised at the same time. Furthermore, the collusion of mobile users augment Sybil attacker's capabilities and significantly reduce the detection accuracy. Therefore, it is crucial to take these challenges into account when developing mobile Sybil detection in MHNs.

In this chapter, we propose a Social-based Mobile Sybil Detection (SMSD) scheme to detect Sybil attackers according to their abnormal contacts and pseudonym changing behaviors in mobile networks. Intuitively, since the Sybil attackers frequently change their pseudonyms to cheat other users, we investigate the contact statistics of the used pseudonyms and detect Sybil attackers by comparing the contact statistics of pseudonyms from normal users and that from Sybil attackers. Due to the limited storage and computation capabilities of mobile users, we utilize cloud servers to store and process the large volume of user's contact information, alleviating the burden of mobile users. The SMSD also addresses the collusion attacks and resists cloud data modification when employing the cloud server for mobile Sybil detection. Specifically, our main contributions are three-fold as follows.

- Firstly, we investigate the characteristics of user's mobile social behaviors, i.e., pseudonym changing and contact. Based on the attacking capabilities, we identify four levels of Sybil attackers, i.e., general Sybil attackers, Sybil attackers with forged contact, Sybil attackers with mobile user's collusion, and Sybil attackers with collusion of cloud servers. Then, we propose a social-based mobile Sybil detection scheme to detect mobile Sybil attackers according to their abnormal pseudonym changing behaviors.

- Secondly, to alleviate the mobile user's storage and computation burden, we exploit the cloud server to store and pre-process the user's contact data. With powerful storage and computation capabilities, the cloud servers can assist to detect the Sybil attackers, which significantly reduces the storage and computation overhead of mobile users.
- Thirdly, we propose a Learning assisted SMSD scheme (LSMSD), i.e., semi-supervised learning with Hidden Markov Model, to resist the collusion of mobile users. The LSMSD can utilize a small number of labeled data for training and adapt to the unlabeled data. We also adopt a ring structure to collect the mobile user's contact signatures associated with the bi-directional Hash chain [11], which can protect user's contact information (e.g., encountered user, contact time, time order, etc.) from being modified by semi-trusted cloud servers.

The remainder of this chapter is organized as follows. We review the related works in Sect. 4.2. The system model, attacker model and design goals are introduced in Sect. 4.3. Then, we present the details of the proposed Sybil detection scheme in Sect. 4.4, followed by security analysis and performance evaluation in Sects. 4.5 and 4.6, respectively. Finally, Sect. 4.7 concludes the chapter.

4.2 Related Works

In this section, we present exiting works to detect Sybil attacks. We review Sybil detection schemes and categorize them into several types, i.e., social network based Sybil detection, social community based Sybil detection, behavior classification based Sybil detection and mobile Sybil detection.

4.2.1 Social Network Based Sybil Detection

Social Network based Sybil detection (SNSD) is a kind of Sybil Detection based on the concept of "social network", which is a social structure linking social relationships among nodes. Sociology theory [12] is a useful tool to investigate the social relationships among users. In this subsection, the term "social network" indicates the user's social graph and structure, which can reflect user's social relationships and the social trustworthiness [13, 14] among users. Leveraging the "social network" structure, Yu et al. [6] propose a famous SNSD scheme, SybilGuard, based on random walk [15, 16]. Before the explanation of the detailed SybilGuard, we give an assumption as follows.

Assumption 1: Although the Sybil nodes can tightly connect with other Sybil ones, the number of social connections among Sybil nodes and honest ones is limited.

SybilGuard relies on Assumption 1, and each node detects the Sybil one in a distributed manner. Specifically, a node with degree R generates totally R random routes starting from itself along its edges with a fixed length L. If a route reaches a known honest node, it is verified by this known honest node. Particularly, a Sybil node S may be accepted as a verified one (i.e., the route from S to H is called verifier) if one of the routes from S reaches the known honest node V. Then, given a threshold $T \leqslant R$, S could be accepted as an honest node when more than T routes from S are verified. With Assumption 1, the limited number of attack edges results in that the number of verifiers cannot be greater than T if T is properly selected. For example, if there are totally X attack edges, the number of Sybil groups is bounded by X. From [17], it is proved that $T = \Theta(\sqrt{n}\log n)$ could be sufficiently large for the honest nodes passing the random walk detection. In addition, security schemes are adopted to ensure the authenticity of the nodes and the routes. Every pair of directly connected two nodes (i.e., one-hop neighbors) negotiate a shared key on the edge. Message Authentication Code (MAC) is used for each node to verify the other one. Furthermore, every generated random route should be registered with an unforgeable token (witness table) including all L nodes on the route so that the attackers cannot deny the connections and forge the route information.

The correctness of SybilGuard relies on the fast-mixing property of the social graph. The mixing time t of a social graph indicates how fast the ending point of a random walk algorithm achieves the stationary distribution. Here, in a social graph, if the ending point distribution is independent on the starting point as $L \longrightarrow \infty$, it is the stationary distribution [6]. If the mixing time is $\Theta(t)$, the graph is fast mixing. When a random walk with the length of $L = \Theta(\sqrt{n}\log n)$, there are $\Theta(\sqrt{n})$ samples that are independent on the starting point. The probability that a Sybil node is accepted by the known honest node (i.e., both the Sybil node and the honest one select the same edge (e.g., attack edge) in the random route) follows the Birthday Paradox [18]. This collision probability is

$$Prob(\text{Collision}) = 1 - \left(1 - \frac{1}{\sqrt{m}}\right)^{\sqrt{m}}. \tag{4.1}$$

Therefore, SybilGuard has a high probability to detect SA-1 according to random walk.

To enhance SybilGuard, Yu et al. [7] propose another defense scheme, called SybilLimit, with the near-optimal guarantees. In SybilLimit, each node generates $R = \Theta(\sqrt{m})$ random routes with length $L = \Theta(\log n)$. Using the random walk algorithm [19], the Sybil or honest nodes can be determined, which is similar to SybilGuard. Different from SybilGuard, SybilLimit leverages the intersections on edges instead of vertex (node), and performs short random routes with multiple independent instances of random walk. SybilLimit accepts $O(\log n)$ Sybil nodes per attack edge, while this number in SybilGuard is $O(\sqrt{n}\log n)$ [7, 20]. Both SybilGuard and SybilLimit are based on Assumption 1.

To understand the properties of social structures, Alvisi et al. [21] investigate several structural properties of social graphs including popularity distribution [22], small world property [12], clustering coefficient [23], conductance [24], etc., and observe that the conductance which is related to the mixing time of a random walk is more resilient in Sybil defense compared with the other properties. Note that popularity distribution among the nodes follows a power-law or lognormal distribution. Small world property indicates that the distance between any two nodes is small. Clustering coefficient is a parameter that reflects the closeness of nodes within a social network. The conductance $C(S)$ reflects the mixing time, which indicates the minimum length of a random walk. $C(S) = \frac{S_{out}}{S_{in}}$, where S_{out} denotes the number of edges that are out from S and S_{in} denotes the number of edges within S. If the conductance is low, the mixing time is high. In [21], it is proved that for the first three properties, the number of edges that Sybil attackers need to generate to launch Sybil attacks is 0 or 1, while this number for the property of conductance is $\frac{C(S)m}{\log(C(S))}$. Sybil attackers have to consume more resources to compete with the conductance based Sybil detection schemes. Therefore, it validates the effectiveness of SybilLimit [7] which utilizes conductance to detect Sybil nodes. In addition, a concept of *perfect attack* is introduced to explain an undetectable attack that draws some honest nodes in the social network into Sybil region, without impact on the whole social network. In other words, when a Sybil node joins the social network and sets up many connections with the honest nodes, it is not easy to detect such an attacker as well. The attack edge is a metric to evaluate the attacker's capability to launch a perfect attack. To resist the strong Sybil attacks, in [21], an SoK defense scheme is proposed exploiting conductance to enable honest users to build a white-list which contains a set of nodes ranked associated with their trustworthiness. The SoK is more robust compared with other SNSD schemes, such as SybilGuard and SybilLimit.

There are many other research efforts on SNSD recently. Cao et al. [25] propose SybilRank to help the centralized Online Social Network (OSN) servers or operators to detect Sybil attacks through ranking nodes according to their perceived likelihood of being Sybils. SybilRank aims to reduce the computation overhead and achieve the scalability of the Sybil detection in a large scale OSN. Danezis et al. [26] explore a probabilistic model of honest node's social network and propose a Bayesian inference approach to divide the whole social graph into Sybil and honest regions. Another Sybil defense [27] adopts the principle of privilege attenuation [28] for SNSD to prevent malicious Sybil attackers adding or removing edges in the social graph without employing social engineering, especially for collusion attack. To further enhance SybilLimit, Tran et al. [29] propose a Sybil detection scheme, Gatekeeper, to achieve optimization for the case of $O(1)$ attack edges and guarantee only $O(1)$ Sybil identities. A multi-source ticket distribution algorithm facilitates Gatekeeper for node admission control.

A state-of-the-art tendency for SNSD is to explore trustworthiness to establish social graph and detect SA-1. SybilFence [30] leverages users' negative feedbacks on Sybil attackers and adjusts the edge weight in the social graph. For example, if a user u_i receives negative comments from others, u_i's edge weights are reduced correspondingly. With the directed social graph, SA-1 can be better detected. Based on credit network [13, 31], SumUp [32] detects SA-1 for the vote aggregation problem in an online content rating system. SumUp leverages online user's voting history in order to restrict the attacker's voting capability if he continuously misbehaves. In SumUp, a trusted node computes a set of max-flow paths on the trusted graph and then aggregates the votes. It allows the votes from the trusted users to be effectively aggregated, while limits the votes from untrusted users. Canal [33] is similar to SumUp. With a credit payment mechanism in a large scale network, Canal enhances the establishment of social graph and is compatible to the existing SNSD. Mohaisen et al. [34] also explore the trust to form the social graph. They rely on the observation that nodes trust themselves more than they trust others, and the trustworthiness on other nodes are not uniformly equal. Then, they leverage differential trust in the social graph to filter weak trust edges and model trustworthiness by biasing random walks. Delaviz et al. [35] propose a trust and credit based Sybil detection scheme, SybilRes, which adopts a local subjective weighted directed graph to indicate user's data transfer activities. When a user u_i uploads data, the edge weight on the path from u_i to the downloaders is reduced. To maintain the edge weight of honest users, after downloading, the downloaders increases the weights of the edges on the paths from the uploader u_i to itself. Then, Sybil users could be detected by using the sophisticated SNSD. Unlike the basic SNSD [6, 7], these trust based SNSD schemes [34, 35] leverage trustworthiness to build a directed social graph rather than the original undirected social graph for random walk Sybil detection. Since this enhancement relies on a practical assumption that the honest nodes would not provide high trust on the unknown (or Sybil) nodes, the attack edges could be filtered to guarantee the SNSD accuracy. Therefore, the credit and trustworthiness enhance the traditional social graph and restrict Sybil attackers to build connections with normal users such that the detection accuracy is improved.

4.2.2 Social Community Based Sybil Detection

Social Community based Sybil Detection (SCSD) explores social community detection to facilitate Sybil detection. The possibility of using social community detection algorithms to detect SA-1 is validated in [36]. In [36], Viswanath et al. first analyze the SNSD schemes and summarizes them to a ranking problem. Since the SNSD schemes usually partition Sybil nodes and honest ones into two parts: Sybil region and honest one, they would be viewed as a graph partitioning problem. For these SNSD schemes, each unknown node is ranked according to its social connections with the known trusted nodes. Then, different parameters (i.e., thresholds) are selected to divide the social graph into two partitions. These parameters determine the boundary of the partition, or "cutoff". The ranking of nodes is towards the direction of reduced conductance. In other words, the nodes tightly connected with the known trusted ones (e.g., lower conductance) would score higher in the ranking. Furthermore, the ranking algorithms significantly impact on the ranking results and the Sybil partition. At the same time, another problem comes out: if a node slightly connects with the current known trusted nodes, it is more likely to be detected as a Sybil node no matter how tightly it connects with other unknown trusted nodes. In other words, when there are multiple social communities in the graph, it is inefficient and ineffective to detect Sybil nodes only through social network partition. Therefore, leveraging community detection to detect Sybil nodes becomes promising and could enhance the Sybil detection accuracy.

SybilDefender [37] is a typical SCSD scheme, which relies on performing a limited number of random walks for Sybil identification and community detection. Sybil identification can detect whether a node is Sybil or not, similar to the existing SNSD schemes. After the Sybil identification, a community detection algorithm is adopted to detect the neighboring Sybil nodes around the detected Sybil one. Furthermore, an efficient combination of Sybil identification and community detection facilitates SybilDefender to further reduce the computation overhead. In addition, due to the observation that a portion of OSN relationships among users are untrusted [22], SybilDefender also includes a mechanism to limit the number of attack edges. This attack edge limiting mechanism enables users to rate their friend's relationships as "Friend" or "Stranger". The attack edges could be further removed since Sybil attackers are probably "Stranger" from the view of normal users. Note that SybilShield relies on Assumption 1.

By using multi-community social network structure, Shi et al. [38] propose Sybil-Shield, an agent-aided SCSD scheme. SybilShield also leverages trust relationships among users to form the social graph. However, due to the fact that two honest nodes belonging to the two different social communities may not tightly connect with each other, SybilShield exploits the agents and ensures the honest nodes tightly connect with other honest ones. Similar to SybilGuard (the first random walk based Sybil detection) [38], some agents of a verifier are selected to run a second round of random walk, called agent walk, where the agents traverse all of the verifier's edges to confirm the suspect nodes. SybilShield relies on Assumption 2.

Assumption 2: Sybil nodes cannot tightly connect honest nodes in the multiple honest communities since honest nodes would not trust Sybil ones. Honest nodes can tightly inter-connect with others in the honest community.

Cai et al. [39] leverage the latent community model and machine learning to detect Sybil attacks, enabling that the tightly interconnected communities are connected more closely than the one loosely connected. Even though some certain communities are compromised by Sybil attackers, the attack communities can also be detected via the transitivity of the latent community model. With a friend invitation graph built according to user's befriending interactions (invite, or accept friends), VoteTrust [40], a novel SCSD scheme, leverages a trust-based vote assignment and global vote aggregation to estimate the probability of a Sybil attacker. VoteTrust combines the social graph structure and user's feedback (accept

Table 4.1 Comparison on social graph based Sybil detection

Sybil defense scheme	Preliminary technique	Social graph	Decentralization	Trustworthiness
SybilGuard and SybilLimit	Random walk	Undirected	$\sqrt{}$	×
SumUp	Adaptive max flow	Undirected	×	Credit network
Gatekeeper	Random walk	Undirected	$\sqrt{}$	Trust
SybilDefender	Community detection	Directed	×	Trust
SybilShield	Community detection	Undirected	$\sqrt{}$	Trust
VoteTrust	Community detection	Directed	$\sqrt{}$	Asymmetric trust

or reject friend requests) to establish a directed graph. It bases on an assumption that the Sybil users cannot receive more than a certain number of friend requests from normal users. The global aggregation of the votes for every node can be used to estimate its global rating. With this two way (voting and feedback) mechanism (e.g., in a directed graph), Sybil detection would be more effective compared others schemes.

In Table 4.1, we compare the SGSD schemes with respect to preliminary techniques, assumption, decentralized properties, etc. A tendency is to explore trustworthiness to facilitate the Sybil detection to SA-1.

4.2.3   Behavior Classification Based Sybil Detection

Users' behaviors can be used to classify Sybil attackers, i.e., Behavior Classification based Sybil Detection (BCSD). In [8, 41], Sybil users in RenRen, a Chinese OSN, can generate an exponential number of social connections with the normal (or honest) users. Jiang et al. [42] also shows that the smarter Sybil attackers rarely establish social connections with other Sybil attackers in RenRen. As a result, only relying on the SGSD schemes cannot effectively detect smarter Sybil attacks since Assumptions 1 and 2 may not always hold. Therefore, some novel Sybil detection schemes are desirable and should exploit some promising features of Sybil attacks.

Wang et al. [41] investigate the OSN user's browsing and clicking habits (as known as "online" habits), to detect the Sybil users by comparing their abnormal online behaviors compared with the normal user's. According to the data obtained from RenRen, the primary OSN activities of users are selected as follows: (1) *Befriending*: send, accept or reject friend requests; (2) *Photo*: upload photos, tag friends in the pictures, browse photos, and comment on the photos; (3) *Profile*: browse profiles of other users; (4) *Share*: share multimedia contents, including video, photo, audio, text contents and website links; (5) *Messaging*: update status, wall posts, send or receive instant messages; (6) *Blog*: post blogs, browse blogs, and comment on the blogs. According to the statistics, the primary activities of

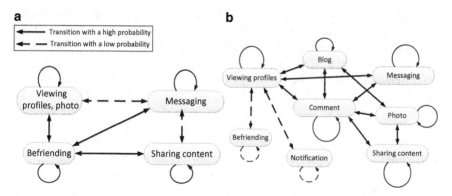

Fig. 4.1 Online social networking behaviors and transition probabilities of Sybil attackers and normal users. (**a**) State transitions for a Sybil user. (**b**) State transitions for a normal user

Sybil users are friending (especially, sending friend requests), viewing photos and profiles of others, and sharing contents with others. On the contrary, the normal users spend a large portion of online time to view photo, and perform other activities, like viewing profiles, sending messages, sharing contents with a similar frequency. Both Sybils and normal users share content or send messages at similar frequencies. Note that sharing content or sending messages are the common approaches for Sybils to disseminate spam in OSNs. This observation indicates that the traditional spam detection schemes cannot simply leverage numeric thresholds to resist spam.

As shown in Fig. 4.1, the click transitions could be modeled by Markov chain with each state as a click pattern. Normal users usually perform diverse OSN behaviors, and the transitions among states are really complicated. By contrast, the Sybil users are involved in some specific activities in a high frequency. To distinguish the BOSAs, Support Vector Machine (SVM) [43, 44] can be adopted according to the session features, such as average clicks per session, average session length, average inter-arrival time between two clicks, and average sessions every day, as well as the click features. The preliminary results show that the Sybil detection accuracy is high. In [41], three models (click sequence model, time based model, hybrid model), which can cluster similar click patterns, are built for the behavior classification. According to some specific similarity metrics, the sequence similarity graph can be established. Through graph clustering, the Sybil users can be detected. The SVM based scheme is supervised learning tool which requires a long term training period. To address this issue, an unsupervised learning scheme is proposed, where only a small portion of click patterns of given normal users as "seeds". They color normal clusters which contain a seed sequence; otherwise, the uncolored clusters are Sybil ones.

With crowdsourcing and social Turing tests, Wang et al. [45] propose a distributed Sybil detection scheme which significantly improves the detection accuracy. For a Sybil attacker, he cannot pass "social Turing test" with different attack strategies. Furthermore, crowdsourcing provides an adaptive platform for normal users (e.g., "turker") to complete the Sybil profile detection with a reasonable cost. From the experiments in [45], the accuracy of crowdsourcing Sybil detection under the reasonable burden is almost as high as that performed by "experts". Some key factors (e.g., demographic factors, temporal factors and survey fatigue, turker selection, and profile difficulty) that may impact on the crowdsourcing Sybil detection are provided. Obviously, the cost of a crowdsourcing workforce is significantly low, which poses a new direction for Sybil detection. In addition, some other BCSD schemes [46] are proposed based on behavior classification. DSybil [46] exploits the heavy-tail distribution of the classical voting behavior from the honest users to detect Sybil identities. In summary, these BCSD schemes can detect SA-2 according to the user's behavior learning and classification.

Suppose that strong Sybil attackers penetrate into the social graph and generate many social connections with the normal users. It opposes the assumption of the SGSD. If Sybil attackers are familiar with normal user's click patterns or habits, i.e., Sybil attackers could truly mimic the normal users, the behavior classification based Sybil detection cannot effectively detect them as well. However, it is obvious that Sybil attackers have to consume a large portion of time to mimic the normal users so that the attack behaviors are partially limited.

4.2.4 Mobile Sybil Defense

In a mobile network, due to the mobility and the lack of global social graph information, Sybil defense is quite different and difficult compared with that in the online networks. Quercia et al. [5] propose an MSD scheme to match mobile user's communities and label the users from the Sybil community as Sybil attackers. In [5], one assumption is that each mobile maintains two lists: friend list containing the trusted mobile users, and foe list with the untrusted users in it. When two users are encountered in the network, they match their communities. If a user is not in the trusted communities, this user would be considered as a Sybil user. In [47], Chang et al. also propose a Sybil defense scheme in Mobile Social Networks (MSNs), assuming that the Sybil users and normal users exist in different communities, and rely on the community matching to detect the Sybil users. Therefore, leveraging friendship is an effective solution to detect Sybil attackers. However, this type of Friend Relationship based Mobile Sybil Detection (FR-MSD) schemes requires mobile users to maintain the trusted community information in advance.

Meanwhile, cryptography is another useful tool to facilitate Sybil defense, especially for Mobile Sybil Defense (MSD), and can restrict Sybil attacker's malicious behaviors. In this subsection, we present some cryptography based MSD (crypto-MSD) schemes based on cryptography techniques to defend mobile Sybil attackers. Vehicular Ad Hoc NETwork (VANET) is one kind of internet of vehicles, characterized by the high-speed mobility. When Sybil attacks are launched in VANET, an added challenge in detecting SA-3 is the mobility that makes it increasingly difficult to tie an attacker to the location. To address Sybil attacks issues in VANETs, Lin [10] proposes an LSR scheme to resist local Sybil attackers and mitigate zero-day Sybil vulnerability in sparse and privacy-preserving VANET. The local vehicle users are not capable to effectively detect Sybil attackers before they are revoked by the TA. To this end, every user u_i should sign on the event that u_i posts. Using group signature [48], if a user sign on the same event for multiple times (e.g., more than one), these signatures may be invalid. Then, the user can be simply linked by other users and detected as Sybil attackers. In [10], the Sybil report delay has been analyzed, while two-layer and multi-layer reporting are introduced to track the Sybil attacker's real identity and for the revocation at the TA's side. Since the pseudonym techniques are widely applied in wireless and mobile networks, there are two sides to the pseudonyms: on one hand, the pseudonym can protect legitimate user's real identity from being identified and linked; on the other hand, pseudonymous identities may hinder the Sybil detection since the detection schemes hardly trace the Sybil identities based on pseudonyms. Similarly, in [49, 50], a malicious user pretending to be multiple (other) vehicles can be detected in a distributed manner through passive overhearing by a set of fixed nodes called road-side boxes. The detection of Sybil attacks in this manner does not require any vehicle in the network to disclose its identity; hence, privacy can be preserved at all times. Triki et al. [51] explore the embedded RFID tags on the vehicles and the short lifetime certificates from RSUs to verify user's authenticity. Some observers (e.g., RSUs, or vehicles) are involved in monitoring the sensitive events to mitigate the false negative detection. Furthermore, vehicles change their identities when they switch to the communication region of another RSU instead of the current one, achieving the unlinkability and privacy.

The advantages of information technologies ensure the service review applications available for the customers using health products and health-related social activities. Users can either use smartphones to query the special offers of health products or features of health-related social events, to browse the reviews or service evaluation from previous customers. Alternatively, local service providers (LSPs) can gather the users' comments and post them to the nearby users via MHNs. Since no trusted authority is available to establish trustworthiness between LSPs and users, Sybil attackers in the local area could forge some positive reviews, delete or modify the negative ones. From the perspective of users, they could also act as an attacker to post fake reviews as well. Therefore, Sybil attacks may maliciously manipulate the system and degrade the quality of smart shopping.

To resist these Sybil attacks, Liang et al. [9] study trustworthy in service evaluation of MSNs, and propose a TSE scheme to facilitate the service review submission and limit the Sybil attackers' capabilities. Specifically, in TSE, LSPs generate plenty of tokens to synchronize mobile users' review submissions. A user u_i can tie his reviews with signatures to only one token after receiving a token from either LSPs or other users having similar profiles or preferences with u_i. The similar profiles and preferences help users build their trust relationships in a local area. Then, the tokens can be circulated among users to enable cooperative review submissions from users with similar profiles or preferences. The efficiency of signature and verification can be achieved with aggregate authentication techniques.

The TSE also embeds the time stamp into the review signature to prevent any user from modifying or deleting the submitted reviews. In addition, every user adopts pseudonym when submitting reviews. All pseudonyms for the reviews in the same token are stored in a list for traceability on a group Sybil attackers. If u_i submits multiple reviews with multiple pseudonyms, both LSP and other users can easily verify it due to the group signature properties. Furthermore, u_i's real identity can be linked due to the revealed multiple pseudonyms that u_i uses. After publishing a token, the LSP cannot omit this token once some reviews are negative to the LSPs. In each token the length of the review chain can bound the LSP's modification capability. For example, the LSP has to be stronger to modify the existing review chain with a longer review chain. With different token structures, such as ring, chain, tree, it is difficult for SA-3 to modify the posted reviews. It is because the established structure would be destroyed if any modification is made on this structure. Besides the basic cryptography solutions, in [9], if a user generates a massive number of reviews with the same pseudonym in a short period, i.e., one time slot, other users can easily detect his behavior.

Some specific features, such as channel characteristics [52, 53] and mobility features, in mobile networks could be investigated to classify Sybil attackers and normal users. For example, in a typical wireless network, channel features are studied to effectively detect Sybil attackers [54]. An enhanced physical-layer authentication is utilized, while the spatial variability of radio channels is typical in indoor and urban environments with rich scattering is exploited. The combination of authentication and channel features detects Sybil attackers. In practice, the proposed scheme is also feasible according to the overhead of the sophisticated channel estimation schemes, either independently or associated with other physical-layer security schemes, like spoofing attack detection. In addition, the received signal strength (RSS) is also used to detect Sybil attackers in a static wireless network, such as wireless sensor networks [55, 56]. If a node always receives the packets with a similar RSS, the sender is probably a Sybil attacker. Some other MSD schemes leverage mobile network features to defend Sybil attacks. Geutte et al. [57] estimate the amount of cheated nodes to measure the success rate of Sybil attacks. They also evaluate the impact of transmission power tuning from senders, while analyze the impact of bi-directional antenna over omni-directional antenna for the receiver. Investigating the transmission signal difference, they quantify the effects of different security assumptions on Sybil attackers and the impact of antennas on the Sybil detection accuracy. Yu et al. [58] also analyze the signal strength distribution of vehicles, and adopt a statistical method to cooperatively verify the location that a vehicle comes from. Since the neighbors cooperatively measure the signal strength of the specific vehicle, the location estimation accuracy can be significantly improved. Abbas et al. [59] propose a lightweight RSS-based Sybil detection scheme in mobile ad hoc network, without centralized authority and dedicated hardware (e.g., directional antenna or GPS). This lightweight detection scheme relies on the node mobility, and sets the threshold to differentiate the node's moving speed. If any node moves much faster than the pre-set threshold, it may be Sybil attackers. In summary, by investigating normal user's and Sybil attacker's behaviors related to the mobility, channel conditions, the SA-3 attackers can be differentiated. The detection strategies would vary in different networks, since the system features also significantly change.

Mobility is an important characteristic of mobile network, and can be adopted to detect Sybil attackers in the mobile environment. Piro et al. [60] observe that in mobile ad hoc network, the Sybil identities related to a single Sybil attacker are bound to a single physical node. In other words, a large number of Sybil identities move together. By monitoring the user's motility, Sybil identities can be detected. Mutaz et al. [61] leverage the features of platoon to detect the Sybils in VANETs. Therefore, defending Sybil attackers through the system features is a promising approach where the challenge is how to obtain the sufficient knowledge or features. Park et al. [62] investigate the mobility of vehicle and rely on the fact that the two vehicles rarely pass multiple roadside units (RSUs) always at the same time. Correlating the vehicles and RSUs in the spatial and temporal domains, Sybil attackers can be identified.

In addition, the secure hardware [63] is used to validate every user's authenticity. Sybil attackers can only authenticate themselves with the a limited number of times, and the fake identities cannot become legal. Although Sybil attacks can be well resisted, the cost of this scheme is huge. Therefore, it would be used in the applications requiring the highest security level. In [64], an identity fee based Sybil defense is proposed, relying on increasing the cost of identity maintenance. The attackers have to spend more fees to launch a Sybil attack. Zhang et al. [65] propose a resource testing scheme to detect the overloaded users which are probably Sybils. The resource testing relies on the observation that the each user or attacker would work on a single or limited number of devices. If a Sybil attacker exists in the network, it might consumes the dramatic amount of resources, such as computation, communication, storage, and network bandwidths, to maintain the created fake identities. Meanwhile, Li et al. [66] propose an admission and retainment control mechanism to enforce nodes to periodically solve computational puzzles. When these dedicated resources can support each node, Sybil attackers would not have adequate recourses to launch the attack. Therefore, the attacker's capabilities are limited to some extend. Reputation systems [67] could be also adopted to mitigate Sybil attacks in mobile network [68, 69]. These Sybil detection schemes provide some challenges, such as hardware, device resource, and reputation, to limit the Sybil attacker's behaviors.

In Table 4.2, we summarize the existing Sybil defense schemes with respect to some design principles. Sybil defense should leverage different features to classify, detect, and resist Sybil attacks towards different scenarios and networks.

In summary, some existing mobile Sybil detection schemes either rely on the pre-defined communities among users, or adopt cryptography techniques to restrict Sybil attackers. However, the Sybil attacker would act similarly as normal users to disrupt mobile Sybil detections. Furthermore, some online Sybil detection schemes cannot be directly applied in the mobile network. To this end, we study the

Table 4.2 Sybil detection: a comparison

Sybil defense scheme	The type of Sybil attacks	Preliminary technique	Base or assumption	Decentralization
SNSD	SA-1	Social graph partition, random walk	Assumption 1	Centralized
SCSD	SA-1	Community detection	Assumption 2	Centralized and decentralized
BCSD	SA-2	Behavior classification	Behavior difference	Centralized and decentralized
FR-MSD	SA-3	Community detection, or profile matching	Trusted community features	Decentralized
Feature-MSD	SA-3	Channel estimation, feature classification	Wireless channel characteristics, mobility features	Decentralized
Cypto-MSD	SA-3	Cryptography	Security of crypto-graphy	Decentralized

relation between mobile user's contacts and pseudonym changing behaviors and propose the mobile Sybil detection scheme balancing the trade-off between the detection accuracy and overhead. Furthermore, the cloud server is introduced into our detection system to alleviate mobile user's resource consumption.

4.3 System Model and Design Goal

In this section, we first introduce the system model and attacker model, and then identify the design goals in details.

4.3.1 System Model

In our system, there are basically three entities, i.e., trusted authority, mobile users and cloud servers as shown in Fig. 4.2.

- **Trusted Authority (TA)** bootstraps the whole system, and generates secret keys to mobile users. Furthermore, the TA audits the mobile users' data stored in cloud servers. After a Sybil attacker is detected, the TA can revoke his identities and update the revocation list.
- **Mobile Users** take smart phones or portable communication devices to bi-directionally communicate with each other. User u_i should first register to

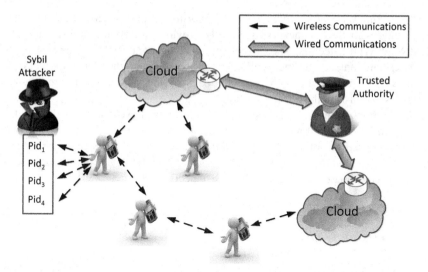

Fig. 4.2 System model

the TA for identity and secret keys which are used to generate session keys, pseudonyms, and signatures. After the registration, u_i adopts pseudonyms to prevent its real identity from being exposed.

- **Cloud Server (CS)** is a semi-trusted entity in the system. The CS has powerful storage and computing capabilities and is deployed in the local area. Furthermore, the CS can directly communicate with mobile users and collect their data.

4.3.2 Attacker Model

According to the Sybil attacker's capabilities, we define Sybil attackers in four levels.

1. *General Sybil Attackers* (Level-1): Sybil attackers, denoted as $\mathscr{A}_s$, exist in mobile environments to compromise the normal users and launch Sybil attacks to maliciously produce biased information to others [70]. $\mathscr{A}_s$ adopts pseudonyms to hide his real identity u_s and repeatedly sends the similar information or spam to normal user u_i. As a result, u_i would consider all the same information from different senders, and u_i's preference may be manipulated by $\mathscr{A}_s$.

2. *Sybil Attackers with Forged Contact* (Level-2): A Sybil attacker $\mathscr{A}_s$ would forge his contact information to benefit $\mathscr{A}_s$ during Sybil detection. In other words, a large number of contact records could provide the evidences of changing pseudonyms. As such, $\mathscr{A}_s$ would maliciously generate an extensive number of fake contact information associated with his pseudonyms to increase the pseudonym changing frequency. Then, $\mathscr{A}_s$ could disrupt the mobile Sybil detection.

3. *Sybil Attackers with Mobile User's Collusion* (Level-3): Mobile users may collude with Sybil attackers to illegally provide fake contact information and disrupt Sybil detection. The colluded users can generate valid signatures on the inexistent contact with the Sybil attackers, even though they have not met each other. As a result, the Sybil attackers can provide valid contact signatures to others and disrupt the Sybil detection.
4. *Sybil Attackers with Collusion of Cloud Servers* (Level-4): Even though the CS can honestly follow the protocol, but it is a semi-trusted entity. If a CS is compromised or colludes with the Sybil attacker $\mathscr{A}_s$, the CS could either add some fake contact information for $\mathscr{A}_s$, or modify and delete the normal user's contact information to increase the false detection rate.

4.3.3 Design Goals

We identify the design goals of Sybil detection as follows.

1. *General Mobile Sybil Detection*: When a Level-1 Sybil attack maliciously changes his pseudonyms to launch attacks, the scheme should detect this from his mobile social behaviors.
2. *Unforgeability*: The proposed scheme should be able to prevent attackers from forging the contact information, since the forged contact would disrupt the Sybil detection. The encountered users should exchange unforgeable information (e.g., signatures of the contact) to the other user, and keep the integrity of contact information.
3. *Resistance to Collusion of Mobile Users*: The proposed scheme should be able to resist the collusion of mobile users. It is critical to find out the forged inexistent contacts when mobile users collude.
4. *Resistance to Collusion of Cloud Servers*: The data stored in the cloud server should not be added, modified or deleted by CSs (i.e., Level-4 attackers). The modified data should be detected by other entities, e.g., the TA and mobile users.

4.4 The SMSD Scheme

In this section, we propose the SMSD scheme to detect the four levels of Sybil attackers. Mobile users collect the contact signature from each encountered user, which is used to support the pseudonym changing, as shown in Fig. 4.3. Collecting the contact information from mobile users, the detector (i.e., the CS) can distinguish Sybil attackers from the normal users according to the abnormal pseudonym changing and contact behaviors. The CS helps to store mobile user's contact

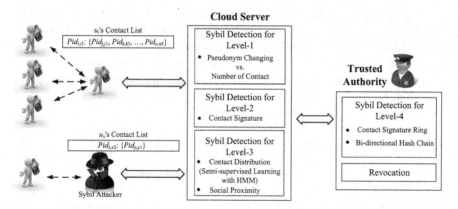

Fig. 4.3 Overview of SMSD

signatures, which considerably reduces their resource consumption. Using semi-supervised learning with Hidden Markov Model (HMM), we propose a novel detection scheme to distinguish the abnormal contact distribution and detect the colluded mobile users.

4.4.1 Social-Based Mobile Sybil Detection

As the pseudonym technique [71] is widely applied in the mobile network, on one hand, it can protect user's real identity from identified and linked by others; on the other hand, the use of pseudonyms may hinder the Sybil detection since mobile users cannot easily link the Sybil identities only based on pseudonyms. Generally, a Sybil attacker $\mathscr{A}_s$ aims to maliciously produce the biased information and convince the normal users. If $\mathscr{A}_s$ uses the same pseudonym to send the same information to user u_i multiple times, u_i would detect them as the spam. If $\mathscr{A}_s$ changes his pseudonyms in a high frequency and sends the same information to u_i multiple times, these information might be originated from different users from the perspective of u_i. As a result, u_i's preference or decision might be impacted by $\mathscr{A}_s$. Therefore, it is of paramount importance to ensure mobile users to legitimately change pseudonyms only if they are encountered many other users.

In general, mobile users adopt period based pseudonym changing (PBPC) strategy and k-anonymity based pseudonym changing (kPC) strategy to protect their anonymity. In the PBPC, a normal user u_i can change his pseudonym after a specific period (or time window) $\mathbb{T}_s$. When using a pseudonym pid_{i,i_p} with a longer duration than $\mathbb{T}_s$, u_i should change it since pid_{i,i_p} is exposed for a long time and might be identified or linked by others. With the PBPC, normal users cannot change their pseudonyms frequently if $\mathbb{T}_s$ is properly selected. The drawback of the PBPC is that

u_i cannot adjust the period according to the number of contacts. Alternatively, the kPC enables the normal user u_i to change his pseudonym pid_{i,i_p} when k-anonymity [71] is violated. In other words, after pid_{i,i_p} is used more than TH times (TH is a pre-defined threshold), pid_{i,i_p} should be changed. Note that it is possible for u_i to change his pseudonym pid_{i,i_p} in a high frequency, if pid_{i,i_p} meets many users (e.g., more than TH users) within a short period. However, u_i would not always change pseudonyms in such a high frequency in reality.

To understand the relation between contact and pseudonym changing behaviors, we investigate the Infocom06 trace [72], which is a real human trace with 78 mobile users attending a conference within 4 days. We collect the contacts and pseudonym changing behaviors from active user (with the highest number of contacts), medium active user (with the average number of contacts), and Sybil attacker (Note that similar to [5], we randomly select users from the trace as attackers, and set the pseudonym changing frequency higher than normal users) in Fig. 4.4. Time is divided into small time slots (i.e., 10 min for each). In Fig. 4.4d, Sybil attackers adopt more pseudonyms under the similar mobility (and contact) of normal users

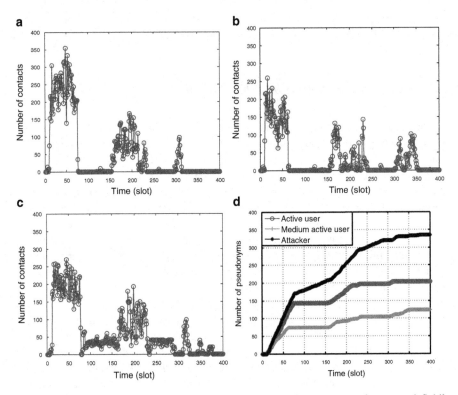

Fig. 4.4 Observations on contact and pseudonym changing between normal users and Sybil attackers. (**a**) Contact of an active user. (**b**) Contact of a medium active user. (**c**) Contact of an attacker. (**d**) Pseudonym changing among normal users and Sybil attackers

Algorithm 3 SMSD

1: **Input:** a mobile user u_i with pseudonym pid_{i,i_p},
   an initialized contact list $\mathscr{CL}_{i,i_p}$, and pseudonym changing threshold TH
2: **Output:** $\mathscr{CL}_{i,i_p}$ and Sybil detection
3: **while** $|\mathscr{CL}_{i,i_p}| < TH$ ($|\mathscr{CL}_{i,i_p}|$ denotes the number of items) **do**
4:    **if** pid_{i,i_p} is encountered with another user pid_{j,j_q} **then**
5:       They generate $\mathbb{C}_{i_p,j_q} = (pid_{i,i_p}, pid_{j,j_q}, t)$.
6:       pid_{i,i_p} adds $\mathbb{C}_{i_p,j_q}$ into $\mathscr{CL}_{i,i_p}$.
7:    **end if**
8: **end while**
9: u_i changes pid_{i,i_p} to $pid_{i,i_{p+1}}$.
10: **Sybil Detection:**
11: Having $\mathscr{CL}_{i,i_p}$, the detector first checks if (1) $|\mathscr{CL}_{i,i_p}| < TH$ and (2) $\mathbb{T}_{p-1} < t_1 < \cdots < t_j < \cdots < t_n < \mathbb{T}_p$. Here, $\mathbb{T}_{p-1}$ and $\mathbb{T}_p$ are starting and ending time of pseudonym pid_{i,i_p}
12: **if** Both (1) and (2) are not guaranteed at the same time **then**
13:    pid_{i,i_p} is maliciously used. u_i is a Sybil attacker.
14: **else**
15:    pid_{i,i_p} is legitimately used.
16: **end if**

---

as shown in Fig. 4.4a, b. Normal users change their pseudonyms in the appropriate way. In contrast, Sybil attackers may sometimes normally change their pseudonyms to act as normal users, and abnormally change their pseudonyms when launching attacks.

In this chapter, we adopt the mobile user's contact information including the encountered user's pseudonym and the number of contacts as the evidence for mobile users to support their pseudonym changing behaviors. The contact between two users with pseudonyms pid_{i,i_p} and pid_{j,j_q} is denoted as $\mathbb{C}_{i_p,j_q} = (pid_{i,i_p}, pid_{j,j_q}, t)$, where t is the encountered time. We utilize kPC as user's pseudonym changing strategy. The detailed Sybil detection steps are shown in Algorithm 3. After the basic Sybil detection, the detector reports the Sybil attacker $\mathscr{A}_s$'s pseudonym and the corresponding contact list to the TA.

4.4.2 Contact Signature with Aggregate Verification

Since the pseudonym pid_{i,i_p} (belonging to $\mathscr{A}_s$) with fewer contacts would be detected as a Sybil attacker, the Level-2 Sybil attacker $\mathscr{A}_s$ may forge the contact information and provide fake contacts to the detector such that $\mathscr{A}_s$ could be merged into the crowd of normal users. To resist Level-2 Sybil attacker, we propose a contact signature scheme. As an evidence of the contact, the contact signature is generated by each pair of the encountered legitimate mobile users. We adopt a variant of aggregate signature [73] to reduce the overall signature size and the verification overhead.

Let $\mathbb{G}$ and $\mathbb{G}_1$ be additive cyclic groups with the same prime order q, and P is the generator of $\mathbb{G}$. $H : \{0,1\}^* \to \mathbb{G}$, and $H_1 : \{0,1\}^* \to \mathbb{Z}_q^*$ are two cryptographic hash functions. Let e be a bilinear pairing, where $e: \mathbb{G} \times \mathbb{G} \to \mathbb{G}_1$ [73] between $\mathbb{G}$ and $\mathbb{G}_1$ exists under two conditions: (1) for any random numbers $a, b \in \mathbb{Z}_q^*$, $e(aP, bP) = e(P,P)^{ab}$; (2) $e(P,P) \neq 1$. Taking a security parameter κ as input, a probabilistic algorithm outputs a tuple $(q, \mathbb{G}_1, \mathbb{G}, e, P, H, H_1)$ as the system parameters to the public.

- **Initialization**: A user u_i receives a series of pseudonyms pid_{i,i_1}, pid_{i,i_2}, ..., pid_{i,i_n}. Each pseudonym pid_{i,i_p} is assigned with the corresponding secret key pair $\mathsf{SK}_{i,i_p} = (sk_{i_p,0}, sk_{i_p,1})$, where $sk_{i_p,0} = s_{i,i_p} H(pid_{i,i_p}||0)$, and $sk_{i_p,1} = s_{i,i_p} H(pid_{i,i_p}||1)$. $s_{i,i_p} \in \mathbb{Z}_q^*$ is selected by u_i. The public key is $\mathsf{PK}_{i,i_p} = s_{i,i_p} P$.
- **Contact Signature**: When two users pid_{i,i_p} (i.e., u_i) and pid_{j,j_q} (i.e., u_j) are encountered, pid_{i,i_p} generates the contact as $\mathbb{C}_{i_p,j_q} = \{pid_{i,i_p}, pid_{j,j_q}, t\}$. pid_{i,i_p}'s signature of the contact between pid_{i,i_p} and pid_{j,j_q} at time t is

$$\mathsf{Sign}_{\mathsf{SK}_{i,i_p}}(\mathbb{C}_{i_p,j_q}) = (pid_{j,j_q}, \omega_{i_p}, \theta_{i_p}) \tag{4.2}$$

$$\begin{cases} \omega_{i_p} = r_{i_p} H(pid_{j,j_q}) + sk_{i_p,0} + c_{i_p} sk_{i_p,1} \\ \theta_{i_p} = r_{i_p} P \end{cases} \tag{4.3}$$

where $c_{i_p} = H_1(t||pid_{i,i_p}||pid_{j,j_q})$, and $r_{i_p} \in \mathbb{Z}_q^*$ is a random number. Finally, pid_{i,i_p} sends $\mathsf{Sign}_{\mathsf{SK}_{i,i_p}}(\mathbb{C}_{i_p,j_q})$ to pid_{j,j_q} as the unforgeable signature to prove the contact $\mathbb{C}_{i_p,j_q}$.
- **Verification**: When receiving the contact signature from the encountered user, pid_{j,j_q} verifies its authenticity as

$$e(\omega_{i_p}, P) \overset{?}{=} e(\theta_{i_p}, H(pid_{j,j_q})) \cdot$$
$$e(H(pid_{i,i_n}||0) + c_{i_p} H(pid_{i,i_n}||1), \mathsf{PK}_{i,i_p}). \tag{4.4}$$

If Eq. 4.4 holds, the received signature is valid; otherwise, it is forged or invalid. Then, pid_{j,j_q} replies $\mathsf{Sign}_{\mathsf{SK}_{j,j_q}}(\mathbb{C}_{i_p,j_q}) = (pid_{i,i_p}, \omega_{j_q}, \theta_{j_q})$ to pid_{i,i_p}. These signatures can be stored and used as the evidence of user's pseudonym changing.
- **Aggregate Authentication**: When pid_{j,j_q} changes his pseudonym to $pid_{j,j_{q+1}}$, u_j collects all the contact signatures related to pid_{j,j_q} and sends them to the CS for Sybil detection. As the increasing number of encountered users, the size of signatures correspondingly increases. To reduce the communication and computation overhead of authentication, we adopt aggregate authentication. First, u_j aggregates the signatures $\mathsf{Sign}_{agg} = (\Omega_{agg}, \Theta_{agg}, pid_{j,j_q})$ of $(pid_{1,1_a}||pid_{2,2_b}|| \ldots ||pid_{i,i_p}|| \ldots ||pid_{n,n_x}, t_1||t_2|| \ldots ||t_i|| \ldots ||t_n, pid_{j,j_q})$ where

$$\Omega_{agg} = \sum_{i=1}^{n} \omega_{i_p}, \Theta_{agg} = \sum_{i=1}^{n} \theta_{i_p}. \tag{4.5}$$

Table 4.3 Comparison of
computation complexity

	Sign	Verification
S	$C_{H_p} + 3C_M$	$3C_{H_p} + 3C_p + 2C_M$
S_{agg}	$N \cdot C_{H_p} + 3N \cdot C_M$	$(2N+1) \cdot C_{H_p} + (N+2) \cdot C_p +$ $(N+1)C_M$

Then, u_j sends the aggregate signature Sign_{agg} to the CS for authentication.

To verify pid_{j,j_q}'s aggregate signature, the CS checks

$$e(\Omega_{agg}, P) \stackrel{?}{=} e(\Theta_{agg}, \mathsf{H}(pid_{j,j_q})) \cdot$$

$$\prod_{i=1}^{N} e(\mathsf{H}(pid_{i,i_p}||0) + c_{i_p}\mathsf{H}(pid_{i,i_p}||1), \mathsf{PK}_{i,i_p}).$$

If it does not hold, some of pid_{j,j_q}'s contact signatures are forged by pid_{j,j_q} or other
mobile users. Note that during each contact, pid_{j,j_q} should check the validity of the
received signatures at the beginning. In other words, every stored contact signature
by pid_{j,j_q} should be valid by pid_{j,j_q}'s verification. The forged signatures would be
forged by pid_{j,j_q}. Therefore, the CS could directly detect the Level-2 Sybil attacker.

The contact signature may increase the communication, computation and storage
overhead, which is crucial to mobile users. We adopt the cloud server to replace
mobile users as the detector. We show the computation complexity in Table 4.3,
where C_{H_p} is map-to-point Hash operation, C_M is multiplication, and C_p is pairing
operation. Our aggregate signature scheme can significantly reduce the verification
overhead. If a mobile user u_i holds all the contact signature, u_i should provide his
historic contact signatures for other user's detection. It would directly expose his
past pseudonyms, while the authentication overhead exponentially increases as u_i
meets more users. In the SMSD, the CS takes u_i's contact signatures and verifies
once for each pseudonym. Then, the CS signs a receipt for the successful detection
to u_i. Thus, u_i can adopt this receipt to prove his completed detection other than
authenticating his past pseudonyms to every user.

4.4.3   Learning Assisted Mobile Sybil Detection

Level-3 Sybil attacker $\mathscr{A}_s$ may collude with other mobile users or attackers to disrupt
the general Sybil detection via maliciously generating valid signatures to prove
the contacts with $\mathscr{A}_s$. These inexistent contacts between $\mathscr{A}_s$ and the colluded users
increase the total number of $\mathscr{A}_s$'s contacts and "validate" his abnormal pseudonym
changing. To this end, we propose Learning assisted SMSD scheme (LSMSD) to
detect Level-3 Sybil attackers to enhance the basic SMSD. Specifically, the LSMSD
basically consists of three steps: contact rate distribution, semi-supervised learning
with HMM and social proximity evaluation.

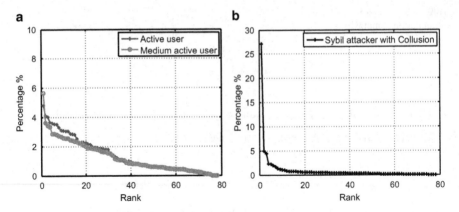

Fig. 4.5 Comparison of contact rate distribution between normal users and Sybil attacker. (**a**) Contact rate distribution of normal users. (**b**) Contact rate distribution of the Sybil attacker

- ***Contact Rate Distribution***: To detect the collusion of mobile users, we first analyze the contact rate distribution with other users. Specifically, when two users are frequently encountered, they are in local vicinity. If two users are colluded, they would have a very high contact rate with each other. Meanwhile, they only have regular or limited contact rate with any other user as shown in Fig. 4.5. The percentage in y-axis is about the contact number for each pair of encountered users. The contact rate distribution could be approximated to an exponential distribution. The detector, e.g., the CS, the TA or other trusted party, can put the contact distribution into sequences for semi-supervised learning with HMM.
- ***Semi-supervised Learning with HMM***: We propose a semi-supervised learning scheme with HMM to detect the collusion of mobile users. First, we utilize an ergodic k-class HMM to analyze the abnormal contact distribution. k is the amount of abnormal states in HMM, while there would be multiple normal states. In the initialization, there is only one state NS_0, which is a central state in HMM as shown in Fig. 4.6. There are l normal states and k abnormal states. NS_0 denotes the basic normal state, and could be obtained by training from a certain number of contact distribution samples. For the ground truth data, we select user's contact distribution during daytime and night time to adjust user's different mobilities and social behaviors.

A set of parameters θ^* of normal state HMM model is obtained from maximizing the likelihood of the ground truth sequence (i.e., contact distribution) $\{c_1^{(l)}, \cdots, c_{N_l}^{(l)}\}$ as

$$\theta^* = \arg\max_{\theta} \prod_{j=1}^{N_l} P(c_j^{(l)} | \theta). \tag{4.6}$$

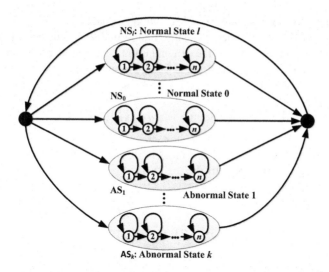

Fig. 4.6 Hidden Markov model

Algorithm 4 LSMSD

1: **Input**: A set of N_l labeled contact distributions $\mathbb{C}^{(l)} = \{c_1^{(l)}, \cdots, c_{N_l}^{(l)}\}$, N_u unlabeled contact distributions $\mathbb{C}^{(u)} = \{c_1^{(u)}, \cdots, c_{N_u}^{(u)}\}$.

2: **Output**: Trained HMM $\Theta = \{\theta_1, \cdots, \theta_K\}$ where $K = l + k$. The classification of the contact distributions $\mathbb{NS} = \{NS_1, \cdots, NS_l\}$ and $\mathbb{AS} = \{AS_1, \cdots, AS_k\}$.

3: Step 1 (**Supervised training**): Given the ground truth data, estimate the central state NS_0 with
$$\theta^* = \arg\max_{\theta} \prod_{j=1}^{N_l} P(c_j^{(l)}|\theta).$$

4: **while** The number of iterations is small than that of abnormal states **do**

5:    Step 2 (**Outlier**): Having a sliding window ω, split the contact distribution into Ω segments with overlapping.
Select the outlier with the smallest likelihood
$$s^* = \arg\min_{s} \left(\arg\max_{s} \prod_{j=1}^{\Omega} P(c_j^{(l)}|s) \right).$$
Label this contact distribution as an abnormal state.

6:    Step 3 (**Adaptation**): A new abnormal state model AS_i is adapted from the general model via the abnormal detection. The normal state model is adapted from the general model by using the other segments.

7:    Step 4 (**Boundary**): Determine the boundary of states

8:    Step 5 (**New Outlier**): Select a new state model with the smallest likelihood in the adaptive normal state model as an outlier.

9: **end while**

We assume that each HMM state follows the Gaussian Mixture Model (GMM), which can be estimated by standard Expectation-Maximization (EM) algorithm [74]. The concrete steps of semi-supervised learning algorithm with HMM are stated in Algorithm 4.

In the adaptation phase, we utilize Maximum A Posteriori (MAP) [75] scheme to adjust the normal state model to a certain abnormal state model for training on the abnormal state model. While, the original normal state model is also trained by adapting the non-outlier segments. We select θ^* to maximize posterior probability density as

$$\theta^* = \arg\max_{\theta} P(\theta|C) = \arg\max_{\theta} P(C|\theta)P(\theta). \qquad (4.7)$$

With GMM, the model is adapted according to the new weight, mean and variance, denoted by w_i', μ_i' and σ_i', respectively.

$$w_i' = \frac{1}{L}\sum_{j=1}^{L} P(i|c_j, \theta), \quad \mu_i' = \frac{\sum_{j=1}^{L} c_j P(i|c_j, \theta)}{\sum_{j=1}^{L} P(i|c_j, \theta)}$$

$$\sigma_i' = \frac{\sum_{j=1}^{N} P(i|c_j, \theta)(c_j - \mu_i')(c_j - \mu_i')^T}{\sum_{j=1}^{L} P(i|c_j, \theta)}. \qquad (4.8)$$

The adaptive parameters can be updated as

$$\hat{w}_i = \beta \cdot w_i + (1 - \beta) \cdot w_i'$$
$$\hat{\mu}_i = \beta \cdot \mu_i + (1 - \beta) \cdot \mu_i'$$
$$\hat{\sigma}_i = \beta \cdot (\sigma_i + (\hat{\mu}_i - \mu_i)(\hat{\mu}_i - \mu_i)^T) \qquad (4.9)$$
$$+ (1 - \beta) \cdot (\sigma_i' + (\hat{\mu}_i - \mu_i')(\hat{\mu}_i - \mu_i')^T).$$

Here, w_i, μ_i and σ_i are the previous weight, mean and variance. Note that β is the adapter factor to balance the new parameters and the previous ones. When β becomes larger, the new parameters would contribute more in the adapted model. In step (4) of Algorithm 4, we follow the sophisticated Viterbi decoding [74] to determine the boundary of states.

The LSMSD requires a small amount of ground truth data for training, which significantly reduces the training overhead and is suitable for mobile Sybil detection. In HMM, we also establish l normal states which leverages from the active users to inactive ones, from the daytime to the night time. With the adaption on the HMM, the LSMSD can improve the learning accuracy with a large amount of unlabeled data.

- **Social Proximity Evaluation**: Although the LSMSD is effective to detect the collusion by classifying the abnormal contact distribution, it might produce false detection if some normal users always stay together. To address this issue, we exploit the social community to facilitate the LSMSD. Actually, if two users frequently contact each other, they would have some specific social relationships, such as colleagues, social friends, neighbors, etc. All these features can be extracted as social communities. Each user u_i maintains a social community vector $\overrightarrow{SC_i} = [1,0,0,\ldots,1,0]$. We define the social proximity $SP_{i,j}$ between u_i and u_j as

$$SP_{i,j} = \frac{|\overrightarrow{SC_i} \cap \overrightarrow{SC_j}|}{|\overrightarrow{SC_i} \cup \overrightarrow{SC_j}|} \in [0,1]. \tag{4.10}$$

According to the detection results in the LSMSD, the user pair (u_i, u_j) with the social proximity $SP_{i,j} < SP$ can be labeled as the Level-3 Sybil attackers. Here, we define SP as the social proximity that normal friends should have. Therefore, the LSMSD can be further enhanced with the social assistance.

4.4.4 Ring Structure of Contact Signature

In the aforementioned subsections, we present the solutions to resist Level-1, Level-2 and Level-3 Sybil attackers via the relation between contacts and pseudonym changing behaviors, aggregate signatures on contacts, and the contact rate distribution, respectively. A condition is that the CS honestly follows all the procedures and cannot be compromised. In reality, the CS is semi-trusted entity as indicated in Sect. 4.3, and is possibly compromised. Therefore, we adopt a ring structure of contact signature to resist the deletion and modification on the contact data sent to the CS.

Before uploading the contact list to the CS, each mobile user should form his contact list in a ring structure where each item cannot be removed or modified by others.

1. u_i first initializes the contact list $\mathscr{CL}_{i_p}$ for pid_{i,i_p}. When u_i begins to use a pseudonym pid_{i,i_p} at time t_0, the contact list is $\mathscr{CL}_{i_p} = \{\mathsf{Sign}_{\mathsf{SK}_{i,i_p}}(\mathbb{C}_{i_p,i_p})\}$, where $\mathbb{C}_{i_p,i_p} = (pid_{i_p,i_p}, pid_{i_p,i_p}, t_0)$.

2. When pid_{i,i_p} meets pid_{j,j_q} at t_1, u_i obtains the contact signature $\mathsf{Sign}_{\mathsf{SK}_{j,j_q}}(\mathbb{C}_{i_p,j_q})$, and updates the contact signature ring as $\mathscr{CL}_{i_p} = \{R_1, \mathsf{Sign}_{\mathsf{SK}_{j,j_q}}(\mathbb{C}_{i_p,j_q})\}$, where $R_1 = (pid_{i,i_p}, t_0, \mathsf{Sign}_{\mathsf{SK}_{i,i_p}}(\mathbb{C}_{i_p,i_p}))$.

   Similarly, when another user pid_{l,l_r} is encountered with pid_{i,i_p} at t_2, pid_{l,l_r} sends the contact signature $\mathsf{Sign}_{\mathsf{SK}_{l,l_r}}(\mathbb{C}_{i_p,l_r})$ to pid_{i,i_p}. pid_{i,i_p} then updates the contact signature ring as $\mathscr{CL}_{i_p} = \{R_1, R_2, \mathsf{Sign}_{\mathsf{SK}_{l,l_r}}(\mathbb{C}_{i_p,l_r})\}$, where $R_2 = (pid_{j,j_q}, t_1, \mathsf{Sign}_{\mathsf{SK}_{i,i_p}}(\mathbb{C}_{i_p,i_p}))$.

3. pid_{i,i_p} recursively builds the ring structure following step (2). When u_i changes pseudonym pid_{i,i_p} at t_N, u_i finalizes the contact signature ring as $\mathscr{CL}_{i_p} = \{R_1, R_2, \ldots, R_N, \mathsf{Sign}_{\mathsf{SK}_{i_p}}(\mathbb{C}'_{i_p,i_p})\}$, where $\mathbb{C}'_{i_p,i_p} = (pid_{i_p,i_p}, pid_{i_p,i_p}, t^*)$ and $t^* = H_1(t_0||t_1||\ldots||t_N)$.

In addition, u_i generates a contact order list $\mathscr{CO}_{i,i_p} = \{CO_0, CO_1, \cdots, CO_N\}$. Let H_2 and $H_3 \colon \{0,1\}^* \to \mathbb{Z}_q^*$ be two cryptographic hash functions. u_i adopts pid_{i,i_p} in the duration $[t_0, t_N]$, and has contacts at $\mathbb{T} = \{t_1, t_2, \cdots, t_{N-1}\}$. For the n-th contact, $CO_n = H_1(h_n \| pid_{j,j_q})$, where pid_{j,j_q} is the encountered user. Here, $h_n = H_1(H_2(h_{n+1} \| t_{n+1}) \oplus H_3(h_{n-1} \| t_{n-1}))$ where $n \in [1, N-1]$. As such, a bi-directional Hash chain is established, where the forward seed is $h_0 = H_1(t_0)$ and backward seed is $h_N = H_1(t_N)$.

The contact signatures form a closed ring, while the established bi-directional Hash chain guarantees the order of every contact time. The contact list should be synchronized with the contact order list to ensure the integrity of the contact information provided by mobile users.

4.5   Security Analysis

In this section, we discuss the security properties of the SMSD scheme according to the defined attacker model in Sect. 4.3.

4.5.1   General Mobile Sybil Detection (Level-1)

The SMSD scheme can resist general Sybil attack since mobile user's contact and pseudonym changing behaviors are correlated. If a Level-1 attacker $\mathscr{A}_s$ launches the attack, i.e., frequently changes pseudonyms, it is difficult for $\mathscr{A}_s$ to collect sufficient contacts to prove the correctness of changing pseudonyms within the short period. The behavior difference between normal users and Level-1 attackers can directly reflect their primary purposes of participating in the mobile network.

Even though $\mathscr{A}_s$ sometimes performs as a normal user and does not change his pseudonym frequently, the other pseudonyms used within a short period or few contacts can also be identified. The SMSD can also restrict Level-1 Sybil attack's malicious behaviors and lead to the higher resource consumption to launch the attack. Thus, it is applicable to practical applications, and performs effective detection in mobile environments.

4.5.2   Contact Unforgeability of Mobile User (Level-2)

Theorem 1. *The SMSD can prevent Level-2 attackers from maliciously forging contacts by using the contact signatures.*

Proof. During the contact between pid_{i,i_p} and pid_{j,j_q}, they sign on the contact event, including the encountered pseudonyms and the time, by using secret keys. Assume that the computational Diffie-Hellman problem in $\mathbb{G}$ is hard. The contact signature $\omega_{i_p} = r_{i_p} H(pid_{j,j_q}) + sk_{i_p,0} + c_{i_p} sk_{i_p,1}$ and $\theta_{i_p} = r_{i_p} P$ are unforgeable, since $sk_{i_p,0}$, $sk_{i_p,1}$ and r_{i_p} are selected by pid_{i,i_p}. Without compromising other users, the forged contact signature ω_{i_p} and θ_{i_p} from $\mathscr{A}_s$ would be detected by the CS, since the CS aggregates $\mathscr{A}_s$'s contact signatures and verifies them at first. Therefore, the contact signature can validate the authenticity of the contact event, which is also the foundation of the SMSD.

4.5.3 Resistance to Collusion of Mobile User (Level-3)

The LSMSD can resist the collusion of mobile users via the semi-supervised learning with HMM on the contact rate distribution. If one mobile user u_i colludes with the Sybil attacker $\mathscr{A}_s$, u_i generates valid signatures for some inexistent contacts with $\mathscr{A}_s$, such that $\mathscr{A}_s$ can change his pseudonyms early prior to the normal changing time point. $\mathscr{A}_s$ and u_i would have a large number of contacts, reflecting a high contact rate in their contact distribution. However, $\mathscr{A}_s$ would meet other users infrequently, such that $\mathscr{A}_s$ has lower contact rates with others. As shown in Fig. 4.5, normal users and Level-3 attackers have different contact distribution. Therefore, the proposed semi-supervised learning with HMM can classify the normal contact distribution and the abnormal one due to the collusion. The detection accuracy will be presented in Sect. 4.6.

As an enhancement of the LSMSD, social proximity is explored to assist the contact distribution. Since the colluded user may not have strong social connections with $\mathscr{A}_s$, but contacts frequently with $\mathscr{A}_s$, the colluded users can be classified according to their social relationships.

4.5.4 Resistance to Collusion of Cloud Server (Level-4)

Theorem 2. *The CS cannot add, modify and remove the mobile user's contacts due to the contact signature ring and bi-directional Hash chain of contact order.*

Proof. Suppose the CS deletes the contact between pid_{i,i_p} and pid_{j,j_q} at t_j. $h_{j-1} \neq H_1(H_2(h_{j+1}||t_{j+1}) \oplus H_3(h_{j-2}||t_{j-2}))$. Similarly, h_{j+1} cannot be recovered as well. As a result, the whole contact order list is invalid. Due to the forward and backward secrecy, the contact order list cannot be forged. If the CS modifies or adds contact signatures for any user, the detectors can find out the CS's malicious operations due to Theorem 1.

In the contact signature ring, if R_2 (e.g., from pid_{j,j_q}) is deleted, t^* cannot be calculated without t_2. Similarly, if the CS adds R_j^* into $\mathscr{CL}_{i_p}$, the contact signature ring cannot be synchronized with the order list. Therefore, the proposed contact signature ring and bi-directional Hash chain can protect the stored contact information from addition, modification and deletion by the CS.

In summary, the SMSD scheme can resist the four levels of Sybil attackers defined in Sect. 4.3.

4.6   Performance Evaluation

In this section, we evaluate the performance of the SMSD based on the trace-driven simulation.

4.6.1   Simulation Setup

We use Infocom06 trace [72] with 78 mobile users during a 4-day conference. Each mobile user carries a dedicated Bluetooth device, which can discover the surrounding users. Totally, 128,979 contacts are recorded. We separate the entire contact data into two parts: 20% of data are the training set to produce mobile users' profiles (e.g., social communities), and the remaining data are used for the simulation.

We assign users with social communities according to the sociology theory. A complete graph G is built up, where each edge $E(u_i, u_j)$ weighted by the total number of contacts between two vertex u_i and u_j. We then refine the graph G with 78 vertices and 2,863 edges by removing the edges weighted less than 100. Based on Bron-Kerbosch algorithm [76], we extract maximal cliques in G. A clique is a complete subgraph where every edge is high-weighted. According to the weight of each maximal clique, we select 100 social communities (i.e., cliques). The selected communities are used for simulation comparison on social connections.

4.6.2   Simulation Results

To show the advantages of SMSD and LSMSD schemes, we compare the detection accuracy with FFL (Friend and Foe list) [5]. In FFL, mobile users detect attackers based on checking their social friend list. Similarly to [5], we randomly select several Sybil attackers from the data set. We select 1, 4, 8, 12, and 16 users as the attackers in our simulation. To quantify the performance, we adopt false positive rate (FPR) and false negative rate (FNR) as the metrics to evaluate the Sybil detection accuracy. A false positive detection results in a normal user being detected as an attacker, while a false negative indicates that a Sybil attacker being regarded as a normal user. We define $\text{FPR} = P_f / (P_f + N) \times 100\%$ where P_f denotes the number of false positive detections, N is the amount of attackers. Similarly, $\text{FNR} = N_f / (N_f + P) \times 100\%$ where N_f denotes the number of false negative detections, P is

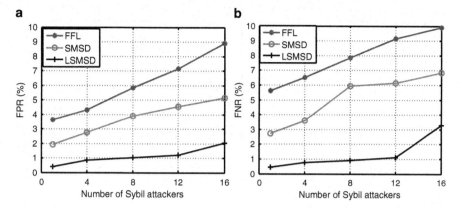

Fig. 4.7 The impacts of the number of Sybil attackers. (**a**) False positive rate vs. the number of Sybil attackers. (**b**) False negative rate vs. the number of Sybil attackers

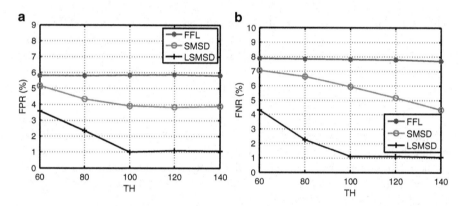

Fig. 4.8 The impacts of TH (i.e., every user changes pseudonyms when the pseudonym meets more than *TH* users). (**a**) False positive rate vs. TH. (**b**) False negative rate vs. TH

the amount of normal users. We run different schemes within 400 time slots for performance comparison. For the LSMSD, we set adapter factor $\beta = 0.5$ and select 100 contact distributions from different users as ground truth data.

In Fig. 4.7, we compare the FPRs and FNRs among FFL, SMSD and LSMSD as the increasing number of attackers. We set $TH = 80$ and $SP = 0.3$. The number of attackers has a larger impact on FFL compared with that on SMSD and LSMSD. The reason is that the increasing number of attackers introduces the larger challenge to the mobile users to detect Sybil attackers via their friend and foe lists. The number can affect SMSD and LSMSD since a large number of Sybil attackers can launch strong collusion attack and forge contact signatures to disrupt the SMSD and LSMSD. In the following results, we set eight Sybil attackers in the network.

In Fig. 4.8, we show the performance comparison by varying the threshold (number of contacts) of changing pseudonyms. We set $SP = 0.3$ for the LSMSD.

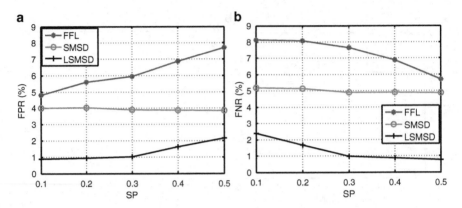

Fig. 4.9 The impacts of SP (i.e., social proximity). (**a**) False positive rate vs. SP. (**b**) False negative rate vs. SP

When *TH* is small, e.g., *TH* = 60, the FPRs and FNRs of SMSD and LSMSD is not high. The reasons are two-fold: on one hand, attackers can also perform as normal users; on the other hand, when attackers launch Sybil attacks, each pseudonym still has a specific lifetime. A smaller *TH* results in a small gap with the number of contacts that an attacker has. When increasing *TH*, the gap becomes larger so that the FPRs and FNRs of SMSD and LSMSD significantly decrease.

As shown in Fig. 4.9, the social proximity threshold *SP* can only impact on LSMSD which detects Level-3 attackers. We set *TH* = 80. For FFL, we adopt *SP* as the threshold to befriend with others. When *SP* is large, the Sybil attackers would not befriend with normal users so that the FNR reduces. Meanwhile, a large *SP* prevents some normal users from befriending with others. Therefore, they may be detected as attackers, which increases the FPR.

For the users with high contact rate, the LSMSD can detect whether their contacts are forged or not according to the social proximity. If *SP* is small, the FNR increases since the colluded attackers may have certain social connections. It is easy to achieve so that both normal users and Sybil attackers with high contact rate are likely detected as normal users. By increasing *SP*, the FNR drops, while the FPR increases. The reason is that the colluded users with high contact rate can hardly build very strong social connections with each other as *SP* > 0.3.

In summary, LSMSD performs better than SMSD since the Level-3 attackers can be detected by semi-supervised learning with HMM, which balances the training overhead and detection accuracy. Having the appropriate parameters, e.g., *TH* = 100, *SP* = 0.3, the four levels of Sybil attackers could be detected.

4.7 Summary

In this chapter, we have proposed a social-based mobile Sybil detection scheme to detect four levels of Sybil attackers with different attacking capabilities. We have investigated mobile user's pseudonym changing behaviors compared with that performed by Sybil attackers, and utilized contact statistics as the criteria of pseudonym changing for mobile Sybil detection. The security analysis demonstrates that the SMSD can resist four levels of Sybil attackers, while the extensive trace based simulation can validate the detection accuracy of the SMSD. The proposed SMSD scheme is a new paradigm in mobile environments, taking the advantages of powerful storage and computing capabilities in the cloud server, and starts a trend to distinguish Sybil attackers via mobile user's contacts and pseudonym changing. The semi-supervised learning with HMM can offer accurate detection with reasonable training overhead. For our future work, we will investigate the cooperation among mobile users to pool the contact statistics for fully distributed Sybil detection.

References

1. K. Zhang, X. Liang, R. Lu, and X. S. Shen, "SAFE: A social based updatable filtering protocol with privacy-preserving in mobile social networks," in *Proc. of IEEE ICC*, 2013, pp. 6045–6049.
2. J. Zhou, Z. Cao, X. Dong, X. Lin, and A. Vasilakos, "Securing m-healthcare social networks: challenges, countermeasures and future directions," *IEEE Wireless Communications*, vol. 20, no. 4, pp. 12–21, 2013.
3. Q. Lian, Z. Zhang, M. Yang, Y. Zhao, Y. Dai, and X. Li, "An empirical study of collusion behavior in the Maze P2P file-sharing system," in *Proc. of IEEE ICDCS*, 2007, pp. 56–66.
4. K. Zhang, X. Liang, R. Lu, and X. Shen, "Exploiting multimedia services in mobile social network from security and privacy perspectives," *IEEE Communications Magazine*, vol. 52, no. 3, pp. 58–65, 2014.
5. D. Quercia and S. Hailes, "Sybil attacks against mobile users: Friends and foes to the rescue," in *Proc. of IEEE INFOCOM*, 2010, pp. 336–340.
6. H. Yu, M. Kaminsky, P. Gibbons, and A. Flaxman, "SybilGuard: Defending against sybil attacks via social networks," *IEEE ACM Transactions on Networking*, vol. 16, no. 3, pp. 576–589, 2008.
7. H. Yu, P. Gibbons, M. Kaminsky, and F. Xiao, "SybilLimit: A near-optimal social network defense against sybil attacks," *IEEE/ACM Transactions on Networking*, vol. 18, no. 3, pp. 885–898, 2010.
8. Z. Yang, C. Wilson, X. Wang, T. Gao, B. Zhao, and Y. Dai, "Uncovering social network sybils in the wild," *CoRR*, 2011.
9. X. Liang, X. Lin, and X. Shen, "Enabling trustworthy service evaluation in service-oriented mobile social networks," *IEEE Transactions on Parallel and Distributed Systems*, vol. 25, no. 2, pp. 310–320, 2014.
10. X. Lin, "LSR: Mitigating zero-day sybil vulnerability in privacy-preserving vehicular peer-to-peer networks," *IEEE Journal on Selected Areas in Communications*, vol. 31, no. 9, pp. 237–246, 2013.

11. K. Zhang, R. Lu, X. Liang, J. Qiao, and X. Shen, "Park: A privacy-preserving aggregation scheme with adaptive key management for smart grid," in *Proc. of IEEE ICCC*, 2013, pp. 236–241.
12. J. Kleinberg, "The small-world phenomenon: An algorithm perspective," in *Proc. of STOC*, 2000, pp. 163–170.
13. A. Cheng and E. Friedman, "Sybilproof reputation mechanisms," in *Proc. of SIGCOMM*, 2005, pp. 128–132.
14. K. Walsh and E. Sirer, "Experience with an object reputation system for peer-to-peer filesharing," in *Proc. of NSDI*, 2006.
15. R. Andersen, F. Chung, and K. Lang, "Local graph partitioning using pagerank vectors," in *Proc. of FOCS*, 2006, pp. 475–486.
16. T. Haveliwala, "Topic-sensitive PageRank: A context-sensitive ranking algorithm for web search," *IEEE Transactions on Knowledge and Data Engineering*, vol. 15, no. 4, pp. 784–796, 2003.
17. R. Morselli, B. Bhattacharjee, M. Marsh, and A. Srinivasan, "Efficient lookup on unstructured topologies," *IEEE Journal on Selected Areas in Communications*, vol. 25, no. 1, pp. 62–72, 2007.
18. P. Flajolet, D. Gardy, and L. Thimonier, "Birthday paradox, coupon collectors, caching algorithms and self-organizing search," *Elsevier Discrete Applied Mathematics*, vol. 39, no. 3, pp. 207–229, 1992.
19. F. Spitzer, *Principles of random walk*. Springer, 1964.
20. H. Yu, "Sybil defenses via social networks: a tutorial and survey," *SIGACT News*, vol. 42, no. 3, pp. 80–101, 2011.
21. L. Alvisi, A. Clement, A. Epasto, S. Lattanzi, and A. Panconesi, "SoK: The evolution of sybil defense via social networks," in *IEEE Symposium on Security and Privacy*, 2013, pp. 382–396.
22. L. Bilge, T. Strufe, D. Balzarotti, and E. Kirda, "All your contacts are belong to us: Automated identity theft attacks on social networks," in *Proc. of WWW*, 2009, pp. 551–560.
23. L. Von Ahn, M. Blum, N. Hopper, and J. Langford, "CAPTCHA: Using hard AI problems for security," pp. 294–311, 2003.
24. J. Leskovec, J. Kleinberg, and C. Faloutsos, "Graphs over time: Densification laws, shrinking diameters and possible explanations," in *Proc. of KDDWS*, 2005, pp. 177–187.
25. Q. Cao, M. Sirivianos, X. Yang, and T. Pregueiro, "Aiding the detection of fake accounts in large scale social online services," in *Proc. of NSDI*, 2012, pp. 1–14.
26. G. Danezis and P. Mittal, "SybilInfer: Detecting sybil nodes using social networks," in *Proc. of NDSS*, 2009, pp. 1–15.
27. P. Fong, "Preventing Sybil Attacks by Privilege Attenuation: A design principle for social network systems," in *IEEE Symposium on Security and Privacy*, 2011, pp. 263–278.
28. P. Denning, "Fault tolerant operating systems," *ACM Computing Surveys*, vol. 8, no. 4, pp. 359–389, 1976.
29. N. Tran, J. Li, L. Subramanian, and S. Chow, "Optimal Sybil-resilient node admission control," in *Proc. of IEEE INFOCOM*, 2011, pp. 3218–3226.
30. Q. Cao and X. Yang, "SybilFence: Improving social-graph-based sybil defenses with user negative feedback," *CoRR*, 2013.
31. P. Dandekar, A. Goel, R. Govindan, and I. Post, "Liquidity in credit networks: A little trust goes a long way," in *Proc. of ACM EC*, 2011, pp. 147–156.
32. D. Tran, B. Min, J. Li, and L. Subramanian, "Sybil-Resilient online content voting," in *Proc. of NSDI*, 2009, pp. 15–28.
33. B. Viswanath, M. Mondal, K. Gummadi, A. Mislove, and A. Post, "Canal: Scaling social network-based Sybil tolerance schemes," in *Proc. of EuroSys*, 2012, pp. 309–322.
34. A. Mohaisen, N. Hopper, and Y. Kim, "Keep your friends close: Incorporating trust into social network-based sybil defenses," in *Proc. of IEEE INFOCOM*, 2011, pp. 1943–1951.
35. R. Delaviz, N. Andrade, J. Pouwelse, and D. Epema, "SybilRes: A sybil-resilient flow-based decentralized reputation mechanism," in *Proc. of IEEE ICDCS*, 2012, pp. 203–213.

36. B. Viswanath, A. Post, K. Gummadi, and A. Mislove, "An analysis of social network-based sybil defenses," in *Proc. of SIGCOMM*, 2010, pp. 363–374.
37. W. Wei, F. Xu, C. Tan, and Q. Li, "SybilDefender: Defend against sybil attacks in large social networks," in *Proc. of IEEE INFOCOM*, 2012, pp. 1951–1959.
38. L. Shi, S. Yu, W. Lou, and T. Hou, "SybilShield: An agent-aided social network-based sybil defense among multiple communities," in *Proc. of IEEE INFOCOM*, 2013, pp. 1034–1042.
39. Z. Cai and C. Jermaine, "The latent community model for detecting sybils in social networks," in *Proc. of NDSS*, 2012, pp. 1–13.
40. J. Xue, Z. Yang, X. Yang, X. Wang, L. Chen, and Y. Dai, "VoteTrust: Leveraging friend invitation graph to defend against social network sybils," in *Proc. of IEEE INFOCOM*, 2013, pp. 2400–2408.
41. G. Wang, T. Konolige, C. Wilson, X. Wang, H. Zheng, and B. Zhao, "You are How You Click: Clickstream analysis for sybil detection," in *Proc. of USENIX*, 2013, pp. 241–255.
42. J. Jiang, C. Wilson, X. Wang, P. Huang, W. Sha, Y. Dai, and B. Zhao, "Understanding latent interactions in online social networks," in *Proc. of IMC*, 2010, pp. 369–382.
43. N. Dalal and B. Triggs, "Histograms of oriented gradients for human detection," in *Proc. of IEEE CVPR*, 2005, pp. 886–893.
44. C. Hsu and C. Lin, "A comparison of methods for multiclass support vector machines," *IEEE Transactions on Neural Networks*, vol. 13, no. 2, pp. 415–425, 2002.
45. G. Wang, M. Mohanlal, C. Wilson, X. Wang, M. Metzger, H. Zheng, and B. Zhao, "Social turing tests: Crowdsourcing sybil detection," in *Proc. of NDSS*, 2012, pp. 1–16.
46. H. Yu, C. Shi, M. Kaminsky, P. Gibbons, and F. Xiao, "DSybil: Optimal sybil-resistance for recommendation systems," in *IEEE Symposium on Security and Privacy*, 2009, pp. 283–298.
47. W. Chang, J. Wu, C. Tan, and F. Li, "Sybil defenses in mobile social networks," in *Proc. of IEEE GLOBECOM*, 2013, pp. 641–646.
48. D. Boneh and H. Shacham, "Group signatures with verifier-local revocation," in *Proc. of CCS*, 2004, pp. 168–177.
49. T. Zhou, R. Choudhury, P. Ning, and K. Chakrabarty, "Privacy-preserving detection of sybil attacks in vehicular ad hoc networks," in *Proc. of MobiQuitous*, 2007, pp. 1–8.
50. ——, "P^{2}DAP - Sybil attacks detection in vehicular ad hoc networks," *IEEE Journal on Selected Areas in Communications*, vol. 29, no. 3, pp. 582–594, 2011.
51. B. Triki, S. Rekhis, M. Chammem, and N. Boudriga, "A privacy preserving solution for the protection against sybil attacks in vehicular ad hoc networks," in *Proc. of WMNC*, 2013, pp. 1–8.
52. K. Ren, H. Su, and Q. Wang, "Secret key generation exploiting channel characteristics in wireless communications," *IEEE Wireless Communications*, vol. 18, no. 4, pp. 6–12, 2011.
53. K. Ren, W. Lou, and Y. Zhang, "LEDS: Providing location-aware end-to-end data security in wireless sensor networks," *IEEE Transactions on Mobile Computing*, vol. 7, no. 5, pp. 585–598, 2008.
54. L. Xiao, L. Greenstein, N. Mandayam, and W. Trappe, "Channel-based detection of sybil attacks in wireless networks," *IEEE Transactions on Information Forensics and Security*, vol. 4, no. 3, pp. 492–503, 2009.
55. M. Demirbas and Y. Song, "An RSSI-based Scheme for Sybil Attack Detection in Wireless Sensor Networks," in *Proc. of WOWMOM*, 2006, pp. 564–570.
56. S. Lv, X. Wang, X. Zhao, and X. Zhou, "Detecting the sybil attack cooperatively in wireless sensor networks," in *Proc. of CIS*, 2008, pp. 442–446.
57. G. Guette and B. Ducourthial, "On the sybil attack detection in VANET," in *Proc. of MASS*, 2007, pp. 1–6.
58. B. Yu, C. Xu, and B. Xiao, "Detecting sybil attacks in VANETs," *Journal of Parallel Distributed Compututing*, vol. 73, no. 6, pp. 746–756, 2013.
59. S. Abbas, M. Merabti, D. Llewellyn-Jones, and K. Kifayat, "Lightweight sybil attack detection in MANETs," *IEEE Systems Journal*, vol. 7, no. 2, pp. 236–248, 2013.
60. C. Piro, C. Shields, and B. Levine, "Detecting the sybil attack in mobile ad hoc networks," in *Proc. of SecureComm*, 2006, pp. 1–11.

61. M. Mutaz, L. Malott, and S. Chellappan, "Leveraging platoon dispersion for sybil detection in vehicular networks," in *Proc. of PST*, 2013, pp. 340–347.
62. S. Park, B. Aslam, D. Turgut, and C. Zou, "Defense against sybil attack in the initial deployment stage of vehicular ad hoc network based on roadside unit support," *Security and Communication Networks*, vol. 6, no. 4, pp. 523–538, 2013.
63. J. Newsome, E. Shi, D. Song, and A. Perrig, "The sybil attack in sensor networks: analysis and defenses," in *Proc. of IPSN*, 2004, pp. 259–268.
64. Y. Reddy, "A game theory approach to detect malicious nodes in wireless sensor networks," in *Proc. of SENSORCOMM*, 2009, pp. 462–468.
65. Q. Zhang, P. Wang, D. Reeves, and P. Ning, "Defending against sybil attacks in sensor networks," in *Proc. of IEEE ICDCS*, 2005, pp. 185–191.
66. F. Li, P. Mittal, M. Caesar, and N. Borisov, "SybilControl: Practical sybil defense with computational puzzles," *CoRR*, vol. abs/1201.2657, 2012.
67. E. Damiani, S. De Capitani di Vimercati, S. Paraboschi, and P. Samarati, "Managing and sharing servents' reputations in P2P systems," *IEEE Transactions on Knowledge and Data Engineering*, vol. 15, no. 4, pp. 840–854, 2003.
68. S. Marti, T. Giuli, K. Lai, and M. Baker, "Mitigating routing misbehavior in mobile ad hoc networks," in *Proc. of MOBICOM*, 2000, pp. 255–265.
69. J. Dinger and H. Hartenstein, "Defending the sybil attack in P2P networks: Taxonomy, challenges, and a proposal for self-registration," in *Proc. of ARES*, 2006, pp. 756–763.
70. K. Zhang, X. Liang, R. Lu, and X. Shen, "Sybil Attacks and Their Defenses in the Internet of Things," *IEEE Internet of Things Journal*, vol. 1, no. 5, pp. 372–383, 2014.
71. X. Liang, X. Li, K. Zhang, R. Lu, X. Lin, and X. Shen, "Fully anonymous profile matching in mobile social networks," *IEEE Journal on Selected Areas in Communications*, vol. 31, no. 9, pp. 641–655, 2013.
72. J. Scott, R. Gass, J. Crowcroft, P. Hui, C. Diot, and A. Chaintreau, "CRAWDAD trace cambridge/haggle/imote/infocom (v. 2006-01-31)," Jan. 2006.
73. D. Boneh, C. Gentry, B. Lynn, and H. Shacham, "Aggregate and verifiably encrypted signatures from bilinear maps," in *Proc. of EUROCRYPT*, 2003.
74. D. Zhang, D. G. Perez, S. Bengio, and I. McCowan, "Semi-supervised adapted HMMs for unusual event detection," in *Proc. of CVPR*, 2005, pp. I: 611–618.
75. D. Reynolds, T. Quatieri, and R. Dunn, "Speaker verification using adapted gaussian mixture models," *Digital Signal Processing*, vol. 10, no. 1–3, pp. 19–41, 2000.
76. C. Bron and J. Kerbosch, "Finding all cliques of an undirected graph (algorithm 457)," *Commun. ACM*, vol. 16, no. 9, pp. 575–576, 1973.

Chapter 5
Privacy-Preserving Health Data Processing

In this chapter, we investigate privacy-preserving health data processing in MHNs to classify health data for diagnosis and prediction with sufficient privacy protection.

5.1 Introduction

With diverse wearable devices and users' healthcare demands, MHNs generate an ever-increasing health data volume. According to an estimation from the Institute for Health Technology Transformation, the United States healthcare data have reached 150 exabytes (1,018 bytes) in 2011 and is expected to reach zettabyte (1,021 bytes) scale and even yottabytes (1,024 bytes) in the future [1]. To support big data in MHNs, health data processing techniques are critical to analyze symptom and diagnose (or predict) disease.

However, as MHNs may take the advantages of the powerful storage and computation capabilities from the outsourced cloud servers, security and privacy concerns associated with these untrusted cloud servers are also raised. When the health data are outsourced to the cloud servers for processing and analysis, the cloud servers can access these raw data. Users' private information might be leaked during this process. It is necessary to make the outsourced health data invisible to the untrusted cloud servers when these cloud servers perform machine learning algorithms [2, 3]. In addition, the user's (e.g., data owner's) identity or the associated profiles should be anonymous or unlinkable.

In this chapter, we investigate health data processing techniques with privacy preservation as follows.

- Firstly, we introduce several basic classification schemes, i.e., hyperplane decision, naive Bayes classification and decision tree, which are able to be extended to other complicated classifications in machine learning. Then, we present Paillier cryptograph system which is also a homomorphic encryption scheme.

© Springer International Publishing Switzerland 2015
K. Zhang, X. Shen, *Security and Privacy for Mobile Healthcare Networks*, Wireless Networks, DOI 10.1007/978-3-319-24717-5_5

It is feasible cryptographic tool to support secure multi-party computations since it can perform some operations (e.g., addition, multiplication) over ciphertexts without learning any information of plaintext.

- Secondly, with the preliminaries, we present the integration of classification schemes and secure computation blocks in a generic privacy-preserving machine learning framework. We present the detailed constructions of important building blocks.

- Finally, we present an application of clinic decision support system with privacy preservation in MHNs. It has an emerging cloud-based e-healthcare system including data owners, cloud server, data processing center, health data center and data requestors.

The remainder of the chapter is organized as follows. We review the related works in Sect. 5.2. We also present several classification schemes in Sect. 5.3. Then, we introduce the preliminaries in Sect. 5.4. We investigate a generic privacy-preserving machine learning framework in Sect. 5.5, including system model, attacker model and building blocks. In addition, we present an MHN application, i.e., privacy-preserving clinic decision support system in Sect. 5.6. Finally, we summarize the chapter in Sect. 5.7.

5.2 Related Works

Machine learning, as a powerful tool for health analysis, has attracted a lot of attentions from both academic and industrial fields. Recently, a major challenge for studies in health-related machine learning is the capture of sufficient ground truth data for training purposes. For example, students have been enlisted to act out activities of daily living (ADLs) to create labeled data sets, which are used to investigate statistical activity recognition methods [4]. A smaller data set from a few volunteers can be also used, for example, mixture model analysis to infer activities of one user, the statistical predictive algorithm to model circadian activity rhythms [5], fuzzy rules used to classify activities in the home [6] and validated with a manual log [7]. Although progress continues, it is still challenging to collect longitudinal sensor data along with real health data of subjects such that the embedded health assessment is also hard.

From the above discussions, it is difficult to obtain labels, but unlabeled data are abundant. It is promising to apply semi-supervised learning, reducing human labor costs but improving learn algorithm accuracy. In some cases, the data are sequential such that most semi-supervised learning methods may not address this problem. Semi-supervised learning can adopt generative and discriminative methods on sequences. A popular generative method is Hidden Markov Model (HMM). Specifically, the standard Expectation-maximization (EM) training with forward-backward algorithm (i.e., Baum-Welch [8]) is a sequence semi-supervised learning algorithm. The training data usually include a small set of labeled data with l labeled sequences $\{\mathbf{X}_L, \mathbf{Y}_L\} = \{(\mathbf{x}_1, \mathbf{y}_1), \cdots, (\mathbf{x}_l, \mathbf{y}_l)\}$, and a much larger unlabeled data set with sequences $\mathbf{X}_U = \{\mathbf{x}_{l+1}, \cdots, \mathbf{x}_{l+u}\}$. $\mathbf{x}_i$ denotes the i-th sequence with length m_i, where $\mathbf{x}_i = (x_{i1}, \cdots, x_{im_i})$. Similarly, $\mathbf{y}_i$ is a sequence of labels $y_{i1} \ldots y_{im_i}$. The initial HMM parameters are estimated by the labeled set. Then, the unlabeled data are used to run the EM algorithm, which improves the HMM likelihood $P(XU)$ to a local maximum. Therefore, the trained HMM parameters are determined by both the labeled and unlabeled sequences. Th mixture models and EM algorithm are paralleled in the i.i.d. case.

The discriminative method is to use a kernel machine for sequences and develop semi-supervised dependency via the kernels. In the recent decade, kernel machines for sequences and complex structures contain Max-Margin Markov networks and Kernel Conditional Random Fields (KCRFs). They can support vector machines for structured data and become the generalization of logistic regression. Although these kernel machines are not specifically designed for semi-supervised learning, we can still use a semi-supervised kernel (e.g., the graph kernels) with the kernel machines. Thus, it makes semi-supervised learning on sequential data applicable [9].

To detect certain or abnormal events, the interested events happen over a small proportion of the total time, such as extractive summarization of raw video events and alarm generation in surveillance systems. The automatic detection of temporal events constitutes a problem in the fields of computer vision and multi-model processing under an umbrella of names (unusual, abnormal or rare events) [10]. Note that such events happen in a low rate or cannot be even anticipated. The abnormal event is usually defined as the event with the rarity, unexpectedness and relevance. For example, they seldom occur, representing the rarity; they may not have been expected in advance, representing unexpectedness; and they are relevant for a particular task, representing relevance. Obviously, abnormal event detection entails various challenges. The rarity of abnormal events indicates that it is not applicable to collect sufficient training data for supervised learning. Furthermore, more than one type of abnormal events may happen in a given data sequence. The event types can be expected to differentiate from another one. Consequently, capturing all abnormal events with training a single model is difficult and exacerbates the problem of learning based on the limited number of data. The unexpectedness of abnormal events indicates that it is difficult to define a complete event lexicon, especially with the consideration of both genre-dependent and task-dependent features of event relevance. Most of related works on event detection focus on specific events, which have prior expert knowledge with well-defined models. Therefore, they are not applicable to detect abnormal events, especially in MHNs for disease diagnosis [11, 12]. Since abnormal events feature with rarity, unexpectedness and relevance and traditional supervised model-based schemes are not applicable, in [13], Zhang et. al. propose a semi-supervised adapted HMM framework. This framework first learns abnormal event models by Bayesian adaptation in an unsupervised pattern from a large amount of (commonly available) training data. It also has an iterative structure and adapts any new abnormal event model at each iteration. Therefore, this framework can address the scarcity of training data and the difficulty in pre-defining abnormal events.

To leverage privacy preservation in machine learning for e-healthcare system, some research efforts have been put in recent years. Graepel et al. [14] adopt a leveled homomorphic encryption scheme [15] to delegate the execution of machine learning algorithm with privacy preservation. Liu et al. [16] propose a privacy-preserving clinical decision support system by using naive Bayesian classification with secure operations. In this clinical decision support system, three types of privacy-preserving health data processing techniques, i.e., naive Bayesian classification, patient disease risk computation and Top-k disease retrieval, are proposed. A k-nearest neighbor classification scheme [17] is proposed with Paillier crypto system [18] to achieve security and privacy goals. To protect the privacy leakage during electrocardiogram (ECG) classification, Barni et. al. [19] propose a privacy-preserving classification scheme based on neural network and linear branching program. Recently, Bost et al. [20] develop a set of secure machine learning classification schemes and propose a library of components. Yuan et al. [21] propose a collaborative learning scheme, which enables each user to encrypt his data and upload the ciphertext to the cloud server. The cloud server performs most of the learning algorithms over these ciphertext without learning the plaintext. A variant of "doubly homomorphic" encryption scheme [22] for secure multi-party computation is adopted to perform flexible operations over the encrypted data.

5.3   Classification Approaches

One of the most important machine learning techniques in MHNs is classification. Formally speaking, a user's input data $\mathbf{x}$ is an l-dimensional vector, i.e., $\mathbf{x} = \{x_1, \cdots, x_l\}$ where $x_i \in \mathbb{R}$ and $i \in \{1, \cdots, l\}$. The classification algorithm $\mathsf{C}(w, \mathbf{x})$: $\mathbb{R}^d \mapsto \{\mathbf{c}_1, \cdots, \mathbf{c}_k\}$ on $\mathbf{x}$ outputs $k^* = \mathsf{C}(w, \mathbf{x}) \in \{1, \cdots, k\}$. k^* is the class to which x corresponds based on the model w.

In this chapter, we mainly present three types of classification schemes, i.e., hyperplane decision based classification, Naive Bayes classification and decision tree.

5.3.1   Hyperplane Decision Based Classification

In hyperplane decision based classification, the model w contains k vectors in $\mathbb{R}^d (w = \{w_i\}_{i=1}^{k})$. The output of this classification is

$$k^* = \arg\max_{i \in [k]} f(w_i, x). \tag{5.1}$$

Here, $f(w_i, x)$ denotes the inner product between w_i and x. It indicates the similarity between the two inputs.

This hyperplane based classification scheme generally works with a hypothesis space H associated with an inner product $f(w,x)$. It can solve a binary classification problem, i.e., $k = 2$. Given two classes c_1 and c_2 An input x is classified in class c_1 if $f(w,m(x)) \geqslant 0$; otherwise it is classified in class c_2. Note that the function $m(\cdot)$: $\mathbb{R}^d \mapsto H$ denotes the feature mapping from $\mathbb{R}^d$ to H. Note that a large class of infinite dimensional spaces can be approximated with a finite dimensional space, such as gaussian kernel. In this example, $m(x) = x$ or $m(x) = xP$, where P is a randomized projection matrix selected in the training phase. In the case with $k \geqslant 2$ classes, we can adopt "one-versus-all" approach for classification. In specific, k different models $\{w_i\}_{i=1}^k$ are trained to discriminate each class from all the others. The decision rule follows Eq. 5.1. It can cover a wide range of common classification algorithms, including SVM, logistic regression, least squares, etc.

5.3.2 Naive Bayes Classification

Naive Bayes classification is another promising classifier. In the model w, each class c_i is associated with a probability $\{\mathsf{Prob}(C = c_i)\}_{i=1}^k$. In the input $\mathbf{x}$, the j-th element x_j is a and occurs in a certain class c_i with a probability $\mathsf{Prob}(X_j = a|C = c_i)$. Here, $a \in D_j$ where D_j is X_j's domain, $j \in \{1, \cdots, d\}$, and $i \in \{1, \cdots, k\}$.

The classifier adopts a maximum a posteriori decision rule to select the class with the highest posterior probability as Eq. 5.2.

$$k^* = \arg\max_{i \in [k]} \mathsf{Prob}(C = c_i | X = x)$$

$$= \arg\max_{i \in [k]} \mathsf{Prob}(C = c_i, X = x) \tag{5.2}$$

$$= \arg\max_{i \in [k]} \mathsf{Prob}(C = c_i, X_1 = x_1, \cdots, X_d = x_d)$$

The second equality holds since the normalizing factor $\mathsf{Prob}(X = x)$ can be omitted with the fixed x. This follows Bayes's rule.

The Naive Bayes model relies on the assumption that $\mathsf{Prob}(C = c_i, X = x)$ can be factorized as

$$\mathsf{Prob}(C = c_i, X_1 = x_1, \cdots, X_d = x_d) = \mathsf{Prob}(C = c_i) \prod_{j=1}^d \mathsf{Prob}(X_j = x_j | C = c_i).$$

$$\tag{5.3}$$

According to Eq. 5.3, each of d features are conditionally indenpendent given the class. For simplicity, we assume that the domain of feature values is discrete and finite such that probabilities $\mathsf{Prob}(X_j = x_j, C = c_i)$ are masses [20].

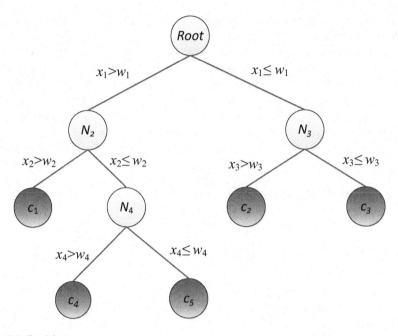

Fig. 5.1 Decision tree

5.3.3 Decision Tree

Decision tree is a non-parametric classifier which partitions the feature vector space one attribute at a time. Interior nodes in the tree correspond to partitioning rules, while leaf nodes correspond to class labels. The decision tree classifies a feature vector **x** starting from the root and uses the partitioning rule at each node to select the next branch until it reaches a leaf node. The output of decision tree based classification [23] is the class associated with the reached leaf node.

Figure 5.1 shows an example of decision tree. Each non-leaf node corresponds to a decision criteria, and each leaf node outputs a certain class $\{c_1, c_2, c_3, c_4\}$.

5.4 Preliminaries

Homomorphic encryption is a useful cryptographic technique to support secure computation among multiple parties. In this section, we introduce the details of Paillier cryptography [18], which is a homomorphic encryption scheme. Paillier cryptography system acts as the building blocks of secure machine learning scheme. Paillier cryptography system consists of key generation, encryption, and decryption.

1. **Key Generation:** An entity chooses two large primes p and q, and computes $N = pq$. The base $\mathbb{G}$ is randomly selected, where $\mathbb{G} \in \mathbb{Z}_{N^2}^*$ and $\gcd(\mathsf{L}(\mathbb{G}^\kappa \bmod N^2)$, $N) = 1$. Here, $\mathsf{L}(x)$ is defined as $(x-1)/N$, while κ is the least common multiple between $p-1$ and $q-1$, $\kappa = lcm(p-1, q-1)$. The public key is $\langle N, \mathbb{G} \rangle$ and the private key is κ.

2. **Encryption:** Let $x \in \mathbb{Z}_N$ be the plaintext and $r \in \mathbb{Z}_N$ be a random number. The ciphertext can be calculated by

$$\mathsf{Enc}(x \bmod N, r \bmod N) = \mathbb{G}^x r^N \bmod N^2 \qquad (5.4)$$

where $\mathsf{Enc}(\cdot)$ is the Paillier encryption operation on two integers modulo N.

3. **Decryption:** Given a ciphertext $c \in \mathbb{Z}_{N^2}^*$, the corresponding plaintext can be derived by

$$\mathsf{Dec}(c) = \frac{\mathsf{L}(c^\kappa \bmod N^2)}{\mathsf{L}(\mathbb{G}^\kappa \bmod N^2)} \bmod N \qquad (5.5)$$

where $\mathsf{Dec}(\cdot)$ denote the decryption operation.

In addition, Paillier cryptography system has a significant property which is especially suitable for multi-parties private computing, since it enables the addition and multiplication over the ciphertext.

Homomorphic: For any $x_1, x_2, r_1, r_2 \in \mathbb{Z}_N^*$, we have

$$\mathsf{Enc}(x_1, r_1)\mathsf{Enc}(x_2, r_2) \equiv \mathsf{Enc}(x_1 + x_2, r_1 r_2) \bmod N^2 \qquad (5.6)$$

$$\mathsf{Enc}^{x_2}(x_1, r_1) \equiv \mathsf{Enc}(x_1 x_2, r_1^{x_2}) \bmod N^2 \qquad (5.7)$$

Self-blinding:

$$\mathsf{Enc}(x_1, r_1) r_2^{x_2} \bmod N^2 \equiv \mathsf{Enc}(x_1, r_1 r_2) \qquad (5.8)$$

5.5   A Generic Privacy-Preserving Machine Learning Framework

A machine learning scheme consists of two phases, i.e., training and classification. In training phase, as shown in Fig. 5.2, the machine learning scheme first learns a model w from the groundtruth data (or a data set of labeled samples and instances). Then, the classification algorithm C classifies over the unseen (or abnormal) feature vector x by using model w to output result or prediction $C(x, w)$.

Usually, a powerful server, such as cloud server, performs the complicated or time-consuming operations. Users input their data to the system and acquire some specific learning results. Due to users' privacy requirements, the feature vector x and the model w should be kept secret during the machine learning. For example, in the

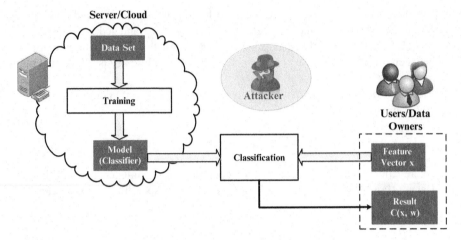

Fig. 5.2 General privacy-preserving machine learning model (The *red boxes* include privacy-sensitive data, and the *white boxes* are operations which should not learn any input or output in plaintext)

hospital-centric e-healthcare system, the patients' health conditions are measured, collected and stored in a remote cloud server due to the large volume of the patients and the related data. However, the cloud server may be untrusted. The patients are not willing to disclose their information to other parties except doctors. Therefore, privacy-preserving machine learning emerges to not only achieve the data storage and processing functionalities but also guarantee users' security and privacy requirements.

5.5.1 Attacker Model

For a general privacy-preserving machine learning scheme, there are passive attackers violating other users' private information (e.g., the raw data, classifier, temporary data, etc.). These attackers honestly follow the communication and computing protocols, but are curious about privacy-sensitive data during machine learning [20]. They try to learn as much private information as possible. Some of this information is sent to attackers, such as cloud servers who help users to perform the complicated operations.

5.5.2 Building Blocks

The generic privacy-preserving machine learning scheme can be achieved by integrating several basic building blocks, such as comparison, argmax, dot product, etc. In this subsection, we present these three building blocks. Three types of cryptograph systems are involved in the building blocks, i.e., Quadratic Residuosity (QR) system [24], Paillier cryptograph system [18] and fully homomorphic encryption (FHE) [25, 26].

Let the plaintext space of QR be $\mathbb{F}_2$ (bits) and $\mathsf{QE}(x)$ be QR encryption of an input bit x. Let the plaintext space of Paillier cryptograph system is $\mathbb{Z}_N$. Here, $N = pq$ is the public modulus where p and q are large primes. Denote $\mathsf{PE}(x)$ as the ciphertext of an integer x under Paillier. For FHE, the plaintext space is $\mathbb{F}_2$. SK_Q and PK_Q are secret and public keys for QR system. Similarly, SK_P and PK_P are Paillier system's secret and public keys.

(a) *Privacy-preserving Comparison (PPC)*

We consider the case that A compares two ciphertexts of integers x and y encrypted by B's public key. The bit length of x and y is l. A holds SK_Q and PK_P. B holds SK_P and PK_Q.

A first computes

$$\mathsf{PE}(a) = \frac{\mathsf{PE}(y) \cdot \mathsf{PE}(l)}{\mathsf{PE}(x)} \bmod N^2$$

and randomly selects $r \in (0, 2^{\lambda+l})$. Then, A computes

$$\mathsf{PE}(\theta) = \mathsf{PE}(a) \cdot \mathsf{PE}(r) \bmod N^2$$

and sends $\mathsf{PE}(\theta)$ to B. B first decrypts $\mathsf{PE}(\theta)$ by using SK_P.

Having random number r, A computes $\omega = r \bmod 2^l$. Meanwhile, B computes $\eta = \theta \bmod 2^l$. Then, A and B privately compute $\mathsf{QE}(t')$ such that $t = 1$ if $\eta < \omega$ according to [27].

A encrypts r_l into $\mathsf{QE}(r_l)$ which is sent to B. Then, B encrypts θ_l and computes $\mathsf{QE}(t) = \mathsf{QE}(t') \cdot \mathsf{QE}(\theta_l) \cdot \mathsf{QE}(r_l)$. $\mathsf{QE}(t)$ is sent to A. Finally, A decrypts t as the comparison result.

The idea of this comparison algorithm is based on the most significant bit of $\theta = y + 2^l - x$ indicates whether $x \leqslant y$.

(b) *Privacy-preserving argmax (PPAM)*

In this case, A wants to output the index of the largest value of $x_1, \cdots, x_n$ which are encrypted under B's secret key. Privacy-preserving argmax algorithm aims to achieve: (1) B can only learn the index of the largest value but learn nothing else. B should not be able to learn the order relations between x_i and x_j. Similar to PPC, A has SK_Q and PK_P, while B has SK_P and PK_Q. PPAM performs as the following steps.

(1) A adopts a random permutation π to prevent B from learning the order of $\{x_1, \cdots, x_n\}$. With π, $\mathsf{PE}(x_i') = \mathsf{PE}(x_{\pi(i)}')$.
(2) A has $\mathsf{PE}(Max) = \mathsf{PE}(x_{\pi(1)})$.

(3) Let $m = 1$. For $i \in (2, \cdots, n)$, B runs PPC with the result b_i in each iteration. $b_i = 1$ if $Max \leqslant a_{\pi(i)}$; otherwise, $b_i = 0$.

(4) In the i-th iteration, A selects two random numbers r_i and $s_i \in (0, 2^{\lambda+l})$. A computes $\mathsf{PE}(m_i') = \mathsf{PE}(Max) \cdot \mathsf{PE}(r_i)$ and $\mathsf{PE}(a_i') = \mathsf{PE}(a_{\pi(i)}) \cdot \mathsf{PE}(s_i)$. Then, $\mathsf{PE}(m_i')$ and $\mathsf{PE}(a_i')$ are sent to B.

(5) B determines if $b_i = 1$ or not, and has

$$\mathsf{PE}(v_i) = \begin{cases} \mathsf{Refresh}(\mathsf{PE}(a_i')), & \text{if } b_i = 1; \\ \mathsf{Refresh}(\mathsf{PE}(m_i')), & \text{otherwise.} \end{cases}$$

Then, B sends $\mathsf{PE}(v_i)$ and $\mathsf{PE}(b_i)$ to A.

(6) A computes

$$\mathsf{PE}(Max) = \mathsf{PE}(v_i) \cdot (g^{-1} \cdot \mathsf{PE}(b_i))^{r_i} \cdot \mathsf{PE}(b_i)^{-s_i}.$$

In plaintext, it indicates $Max = v_i + (b_i - 1)r_i - b_i t_i$.

(7) After n iterations, B sends m to A. Finally, A has the result $\pi^{-1}(m)$.

Note that in each iteration, B can randomize the encryption after determining the maximum of the compared two values. In [20], a "refresh" algorithm is proposed to randomize Paillier ciphertexts. If the "refresher" knows the secret key, it decrypt the ciphertext and re-encrypts it; otherwise, it multiplies a ciphertext of 0.

(c) *Privacy-preserving Dot Product (PPDP)*

In MHNs, the measured health data are usually compared with samples or health database to evaluate if any potential disease on users or patients. It is necessary to enable privacy-preserving dot product over encrypted data. With the homomorphic properties of Paillier cryptograph system, it is feasible to achieve this goal.

When A has a vector $\mathbf{x} = (x_1, \cdots, x_l)$ and B has a vector $\mathbf{y} = (y_1, \cdots, y_l)$ where $x_i, y_i \in \mathbb{Z}, i \in (1, \cdots, l)$. In addition, A has PK_P and B has SK_P. The detailed steps of PPDP are as follows.

(1) B first encrypts $\mathbf{y}$ into ciphertexts as $\mathsf{PE}(y_1), \cdots, \mathsf{PE}(y_l)$ which are sent to A.

(2) A computes $\mathsf{PE}(v) = \prod_{i=1}^{l} \mathsf{PE}(y_i)^{x_i} \bmod N^2$ after receiving ciphertexts from B.

PPC, PPAM and PPDP algorithms are all secure under honest-but-curious model if Quadratic Residuosity cryptograph system and Paillier cryptograph system are semantically secure.

In summary, privacy-preserving classifiers can be performed with the provided building blocks. The security of these Privacy-preserving classifications holds as the building blocks are secure under honest-but-curious model.

5.6 Patient-Centric Clinical Decision Support System

In this section, we present an MHN application, i.e., patient-centric clinical decision support system. It utilizes naive Bayesian classifier to make clinical decision in a real-time manner, which improves the diagnosis accuracy and reduces patients' waiting time.

5.6.1 System Model

In this system, there are six entities, i.e., trusted authority (TA), data owners, cloud server (CS), health data center (HDC), data processing center (DPC) and data requestor (DR) as shown in Fig. 5.3.

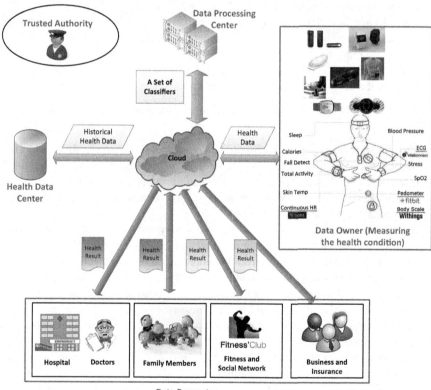

Fig. 5.3 System model of patient-centric clinic decision support system

- **Trusted Authority (TA)** can bootstrap the system and generates secret key materials to all the legal entities. When a user or entity behaves maliciously, the TA revokes their identities and updates the revocation list.
- **Data Owner** can be patients or users who measure their physiology parameters via e-healthcare system. Users encrypt and send the measured health related information to the cloud server. A user u_i first registers to the TA for identity and secret key materials.
- **Cloud Server (CS)** is an untrusted entity in the system. The CS has powerful computational and storage capabilities. CS can perform the complicated and time-consuming operations over health data to release the burden from data owners.
- **Health Data Center (HDC)** stores all the historical health data from users or hospitals. HDC can provide these data in ciphertext for training.
- **Data Processing Center (DPC)** can be data analytic agent or company who receives outsourced professional health data processing requests. The data shared to the data processing center is in ciphertext, while the classifiers from data processing center are also protected from leaking to other parties.
- **Data Requestor (DR)** includes a wide range of third parties, e.g., hospital, doctors, users' family members of patients and insurance company. The DR sends data requests to the CS with different purposes. For example, doctors may acquire real-time health condition of his patients, while insurance company wants to know if a customer has some certain phenomena.

The work flow of this patient-centric clinic decision support system is as follows.

- *Health Data Measurement*: In this phase, data owners utilize on-body sensors and wearable devices to measure their physiology parameters, such as temperature, heart rate, ECG, etc. Since these health data are highly privacy-sensitive to the data owners, it is necessary to encrypt the measured data to ciphertext. Then, the encrypted data are sent to the CS.
- *Health Evaluation Request*: Different users may have different requests with various purposes. For example, the hospital and doctors may need the continuous and real-time health condition from patients (a type of data owners) for analysis and diagnosis. Patient and his family members may want the diagnosis report instead of the complicated health parameters. For fitness club members, only the comparison of several social friends' results is required. For business use, a health insurance company may want the prediction on the measured results.
- *Health Data Processing*: The CS performs the operations as follows: (1) train the historical data from the HDC with the model or classifier from the DPC; (2) classify the new measured health data from data owners based on the model and classifier; (3) distribute the desired results to different data requestors.

5.6.2   Attacker Model and Design Goals

In the attacker model, we define DPC, DR, CS can honestly follow the protocols but are curious about other entities' data (honest-but-curious).

We identify the design goals in terms of computational efficiency and security. The goal is to achieve the secure classification over encrypted health data and distribute diverse data processing results to different requestors without privacy leakage. Data owner's health data should be protected from disclosing to other unauthorized entities. Historical data should also be protected during the training and other operations by other entities. The HDC's processing components, such as classifier, should be protected from studying by other untrusted entities. DR's results should be invisible to other data requestors and untrusted entities in the system.

5.6.3   Privacy-Preserving Clinic Decision Support System

In [16], a novel privacy-preserving clinic decision support system (PPCD) is proposed to achieve privacy-preserving machine learning [28] in e-healthcare system. The PPCD can predict patients' disease risk with three components: (a) privacy-preserving health data training for naive Bayesian classifier, (b) privacy-preserving disease evaluation, and (c) privacy-preserving retrieval of top-k diagnosed results.

(a) *Privacy-preserving Health Data Training for Naive Bayesian Classifier*

The goal of the first component is to acquire the historical health data from HDC and enable DPC to train naive Bayesian classifier. First, these data are sent to CS. Then, CS aggregates and sends data to DPC for training.

(1) Key Generation: TA runs the key generation algorithm to generate public and private keys for HDC and DPC. TA first chooses two cyclic group $\mathbb{G}_1$ and $\mathbb{G}_2$, where N is the prime order and g is the generator of $\mathbb{G}_1$. HDC i selects random numbers $a_{i,1}$ and $a_{i,2} \in \mathbb{Z}_N$. HDC i 's public key is $\mathsf{PK}_i = \{e(g,g)^{a_{i,1}}, g^{a_{i,2}}\}$ and secret key is $\mathsf{SK}_i = \{a_{i,1}, a_{i,2}\}$. Similarly, DPC's public key is $\mathsf{PK}_{DPC} = \{e(g,g)^{p_1}, g^{p_2}\}$ and secret key is $\mathsf{SK}_{DPC} = \{p_1, p_2\}$.

(2) Re-key Generation: TA generates the re-encryption key, which is derived from HDC i's secret key a_i, DPC's public key g^{p_2} and a random number $r'_i \in \mathbb{Z}_N$. $\mathsf{SK}_{i \to DPC} = (g^{p_2})^{a_{i,1}} g^{r'_i}$. In addition, another secret key $S = e(g,g)^{\sum r_i r'_i p_2^{-1}}$.

(3) Encryption: HDC i encrypts the data $x^{(i)} \in \mathbb{Z}_N$ by using PK_{a_i}. The ciphertext is $Enc_{\mathsf{PK}_i}(x^{(i)}) = \{C_{i,1}, C_{i,2}\}$, where $C_{i,1} = g^{r_i}$ and $C_{i,2} = e(g,g)^{x^{(i)}} \cdot e(g,g)^{a_{i,1}r_i}$. Here, $r_i \in \mathbb{Z}_N$.

(4) Decryption: HDC decrypts $Enc_{\mathsf{PK}_i}(x^{(i)})$ by

$$\frac{C_{i,1}}{e(g, C_{i,1})^{a_{i,1}}} = \frac{e(g,g)^{x^{(i)}} \cdot e(g,g)^{a_{i,1}r_i}}{e(g, g^{r_i})^{a_{i,1}}} \tag{5.9}$$

$$= e(g,g)^{x^{(i)}}.$$

As $x^{(i)}$ is considered as a small number in a finite set, HDC can compute discrete logarithm of $e(g,g)^{x^{(i)}}$ and obtain $x^{(i)}$ [22, 29].

(5) Re-encryption with aggregation: CS re-encrypts and aggregates the data $x^{(i)}$. First, $Enc_{PK_i}(x^{(i)})$ is re-encrypted into DPC's domain by using $SK_{i \to DPC}$. The re-encrypted ciphertext is

$$
\begin{aligned}
Enc_{PK_{DPC}}(x^{(i)}) &= \{D_{i,1}, D_{i,2}\} \\
&= \{e(g,g)^{a_{i,1}p_2 r_i} \cdot e(g,g)^{r_i' r_i}, e(g,g)^{x^{(i)}} \cdot e(g,g)^{a_{i,1} r_i}\}.
\end{aligned}
\tag{5.10}
$$

Here,

$$
\begin{aligned}
D_{i,1} &= e(SK_{i \to DPC}, C_1) \\
&= e(g^{a_{i,1}p_2} g^{r_i'}, g^{r_i}) \\
&= e(g,g)^{a_{i,1}p_2 r_i} \cdot e(g,g)^{r_i' r_i}.
\end{aligned}
\tag{5.11}
$$

Then, CS aggregates the l ciphertexts as $C_{agg} = \{C_1', C_2'\}$, where

$$
C_1' = \prod_{i=1}^{l} D_{i,1} = e(g,g)^{p_2 \sum_{i=1}^{l} a_{i,1} r_i} \cdot e(g,g)^{\sum_{i=1}^{l} r_i' r_i}
$$

and

$$
C_2' = \prod_{i=1}^{l} D_{i,2} = e(g,g)^{\sum_{i=1}^{l} x^{(i)}} \cdot e(g,g)^{\sum_{i=1}^{l} a_{i,1} r_i}.
$$

Note that l is the number of HDCs, and $i \in \{1, \cdots, l\}$. This re-encryption and aggregation algorithm is named *ReAgg*.

(6) Re-decryption: DPC decrypts the aggregated ciphertext C_{agg} with SK_{DPC} as

$$
\begin{aligned}
C_2' \cdot S \cdot (C')_1^{-\frac{1}{p_2}} &= \frac{e(g,g)^{\sum_{i=1}^{l} x(i)} \cdot e(g,g)^{\sum a_{i,1} r_i} \cdot e(g,g)^{\sum r_i r_i' p_2^{-1}}}{e(g,g)^{\sum a_{i,1} r_i} \cdot e(g,g)^{\sum r_i r_i' p_2^{-1}}} \\
&= e(g,g)^{\sum_{i=1}^{l} x(i)}.
\end{aligned}
\tag{5.12}
$$

$\sum_{i=1}^{l} x(i)$ is computed with discrete logarithm since it is a small number in a finite set.

Now, we present how to train naive Bayesian classifier with privacy preservation. $\mathbf{X}_i = \{X_{i,1}, \cdots, X_{i,m}\}$ denotes data owner (e.g., patient, user) i's symptom vector, and $\mathbf{Y}_i = \{Y_{i,1}, \cdots, Y_{i,n}\}$ denotes i's disease vector. m and n are the number of symptom categories and the number of disease categories. $X_{i,j} = 1$ if i has the j-th symptom; $X_{i,j} = 0$ otherwise. Similarly, $Y_{i,j} = 1$ if i has the j-th disease; $X_{i,j} = 0$.

HDC t encrypts the historical symptom data $Enc_{\mathsf{PK}_t}(X_{i,j})$ and the corresponding disease data $Enc_{\mathsf{PK}_t}(Y_{i,j})$ with his public key PK_t. These ciphertexts are sent to CS for storage. Then, CS re-encrypts the ciphertexts $Enc_{\mathsf{PK}_t}(X_{i,j})$ and $Enc_{\mathsf{PK}_t}(Y_{i,j})$ from HDC's domain to DPC's domain, i.e., $Enc_{\mathsf{PK}_{DPC}}(X_{i,j})$ and $Enc_{\mathsf{PK}_{DPC}}(Y_{i,j})$. CS also aggregates the symptom vector

$$Enc_{\mathsf{PK}_t}(\mathbf{X}_i) = (Enc_{\mathsf{PK}_t}(X_{1,i}), \cdots, Enc_{\mathsf{PK}_t}(X_{1,m}))$$

into another encrypted vector

$$Enc_{\mathsf{PK}_t}(\mathbf{X}_i') = (Enc_{\mathsf{PK}_t}(X_{i,1}'), \cdots, Enc_{\mathsf{PK}_t}(X_{i,m}'))$$

where

$$Enc_{\mathsf{PK}_t}(X_j') = ReAgg(Enc_{\mathsf{PK}_t}(X_{1,j}), \cdots, Enc_{\mathsf{PK}_t}(X_{m,j})).$$

Note that l is the number of historical health data from HDC. Similarly, CS aggregates the disease vector

$$Enc_{\mathsf{PK}_t}(\mathbf{Y}_i) = (Enc_{\mathsf{PK}_t}(Y_{1,i}), \cdots, Enc_{\mathsf{PK}_t}(Y_{1,n}))$$

into another encrypted vector

$$Enc_{\mathsf{PK}_t}(\mathbf{Y}_i') = (Enc_{\mathsf{PK}_t}(Y_{i,1}'), \cdots, Enc_{\mathsf{PK}_t}(Y_{i,n}'))$$

where

$$Enc_{\mathsf{PK}_t}(Y_j') = ReAgg(Enc_{\mathsf{PK}_t}(Y_{1,j}), \cdots, Enc_{\mathsf{PK}_t}(Y_{n,j})).$$

The aggregated data are sent to DPC. DPC uses his secret key SK_{DPC} to decrypt these ciphertexts and have the vectors.

Then, DPC computes

$$\mathsf{Prob}(A_j = 1 | C_t = 1) = \frac{X_j'}{Y_t'},$$

$$\mathsf{Prob}(A_j = 1 | C_t = 0) = \frac{X_j'}{l - Y_t'},$$

$$\mathsf{Prob}(C_t = 1) = \frac{Y_t'}{l},$$

$$\mathsf{Prob}(C_t = 0) = \frac{l - Y_t'}{l},$$

$$\mathsf{Prob}(A_j = 0 | C_t = 1) = 1 - \mathsf{Prob}(A_j = 1 | C_t = 1),$$

$$\mathsf{Prob}(A_j = 0 | C_t = 0) = 1 - \mathsf{Prob}(A_j = 1 | C_t = 0).$$

$\mathsf{Prob}(A_j | C_t)$ where $j = 1, \cdots, m$ and $t = 1, \cdots, n$ can be used to classify symptom and disease.

However, these probabilities cannot be directly encrypted since the above cryptograph scheme can only encrypt integers. Before encryption, the probabilities should be expanded to integers as $\mathsf{Prob}(A_j) = \lfloor \mathsf{Prob}(A_j) \cdot I \rfloor$, where I is the expansion factor. Usually, I can be 100, 1000, etc.

(b) *Privacy-preserving Disease Evaluation*

When a user or patient (i.e., data owner) measures his health information via wearable devices, the measured data can be sent via MHNs which provides diagnosis results. User u has m symptoms $X_1, \cdots, X_m$ which are encrypted as $Enc_{\mathsf{PK}_u}(X_1), \cdots, Enc_{\mathsf{PK}_u}(X_m)$. These ciphertexts are sent to DPC. Since DPC does not have u's secret key, DPC cannot directly decrypt u's symptom from ciphertexts. According to Bayes theory [30], DPC can compute the user's probabilities of disease. For a disease t, DPC computes

$$Enc_{\mathsf{PK}_u}(\mathsf{Prob}(A_j = X_j | C_t = 1))$$

$$= Enc_{\mathsf{PK}_u}(X_j \mathsf{Prob}(A_j = |C_t = 1) + (1 - X_j)\mathsf{Prob}(A_j = 0 | C_t = 1))$$

$$= Enc_{\mathsf{PK}_u}(X_j)^{\mathsf{Prob}(A_j = |C_t = 1)} \cdot Enc_{\mathsf{PK}_u}(1 - X_j)^{\mathsf{Prob}(A_j = 0 | C_t = 1)}.$$

Similarly, DPC can compute $Enc_{\mathsf{PK}_u}(\mathsf{Prob}(A_j = X_j | C_t = 0))$ and aggregate

$$K_{t,1} = Enc_{\mathsf{PK}_u}(\prod_{j=1}^{m} \mathsf{Prob}(A_j = X_j | C_t = 1) \cdot \mathsf{Prob}(C_t = 1)).$$

Let

$$H_{t,1} = \prod_{j=1}^{m} \mathsf{Prob}(A_j = X_j | C_t = 1) \cdot \mathsf{Prob}(C_t = 1),$$

we have $K_{t,1} = Enc_{\mathsf{PK}_u}(H_{t,1})$. Then, DPC encrypts disease name DN_t with u's public key and outputs the ciphertext $DS_{t,1} = Enc_{\mathsf{PK}_u}(DN_t)$. Similarly, DPC computes $K_{t,0}$ and $DS_{t,0}$. We have

$$K_{t,0} = Enc_{\mathsf{PK}_u}(H_{t,0}) = Enc_{\mathsf{PK}_u}(\prod_{j=1}^{n} \mathsf{Prob}(A_j = X_j | C_t = 0) \cdot \mathsf{Prob}(C_t = 0)),$$

$$DS_{t,0} = Enc_{\mathsf{PK}_u}(DN_{t,0}).$$

Here, $DN_{t,0} = 0$.

If an attacker knows the order of disease, it is possible to violate user's privacy. To address this issue, in [16], a permutation of all the ciphertexts is proposed. In specific, DPC first randomly permutes $K_{t,\pi_1(e)}$ and $DS_{t,\pi_1(e)}$ where $e \in \{0,1\}$. with the permutation π_1. With another permutation π_0, DPC permutes $K_{\pi_0(t),\pi_1(e)}$ and $DS_{\pi_0(t),\pi_1(e)}$ where $e \in \{0,1\}$ and $t \in \{1, \cdots, n\}$. Then, $\pi_0(t)$ and $\pi_1(e)$ are securely stored by CS.

(c) *Privacy-preserving Top-k Diagnosis*

In this phase, the goal is to retrieve the top-k diagnosis results based on the evaluation. Let $P_a = Enc_{\mathsf{PK}_u}(DN_{a,1})$ if $H_{a,1} \geqslant H_{a,0}$; $P_a = Enc_{\mathsf{PK}_u}(DN_{a,0})$ where $a = \pi_0(t)$. Then, CS selects $P'_1, \cdots, P'_k$ from $P_1, \cdots, P_n$ where $H'_1, \cdots, H'_k$ are top-k probabilities of $H_1, \cdots, H_n$. As discussed before, a variant of privacy-preserving argmax scheme can achieve this goal.

In summary, the clinic decision support system can perform health data classification over the encrypted data and preserve privacy of each entity in MHNs.

5.7   Summary

In this chapter, we have investigated privacy-preserving health data processing schemes. We have introduced several basic classification schemes and homomorphic encryptions as preliminaries. Then, we have presented the integration of classification schemes and secure computation blocks. Finally, we have discussed an application of clinic decision support system with privacy preservation.

References

1. "iHT^2 Releases Big Data Research Report," 2013. [Online]. Available: http://ihealthtran.com/wordpress/2013/03/iht2-releases-big-data-research-report-download-today/
2. M. Barreno, B. Nelson, R. Sears, A. Joseph, and J. Tygar, "Can Machine Learning Be Secure?" in *Proc. of ASIACCS*, 2006, pp. 16–25.
3. M. Wen, R. Lu, K. Zhang, J. Lei, X. Liang, and X. Shen, "PaRQ: A Privacy-Preserving Range Query Scheme Over Encrypted Metering Data for Smart Grid," *IEEE Transactions on Emerging Topics in Computing*, vol. 1, no. 1, pp. 178–191, 2013.
4. G. Singla, D. J. Cook, and M. Schmitter-Edgecombe, "Recognizing independent and joint activities among multiple residents in smart environments," *J. Ambient Intelligence and Humanized Computing*, vol. 1, no. 1, pp. 57–63, 2010.
5. G. Virone, M. Alwan, S. Dalal, S. W. Kell, B. Turner, J. A. Stankovic, and R. A. Felder, "Behavioral patterns of older adults in assisted living," *IEEE Transactions on Information Technology in Biomedicine*, vol. 12, no. 3, pp. 387–398, 2008.
6. B. A. Majeed and S. J. Brown, "Developing a well-being monitoring system - modeling and data analysis techniques," *Appl. Soft Comput*, vol. 6, no. 4, pp. 384–393, 2006.
7. T. S. Barger, D. E. Brown, and M. Alwan, "Health-status monitoring through analysis of behavioral patterns," *IEEE Transactions on Systems, Man, and Cybernetics, Part A*, vol. 35, no. 1, pp. 22–27, 2005.
8. L. Rabiner, "A Tutorial on Hidden Markov Models and Selected Applications in Speech Recognition," *Proceedings of the IEEE*, vol. 77, no. 2, pp. 257–286, 1989.
9. Z. Xing, J. Pei, and E. J. Keogh, "A brief survey on sequence classification," *SIGKDD Explorations*, vol. 12, no. 1, pp. 40–48, 2010.
10. M. T. Chan, A. J. Hoogs, J. Schmiederer, and M. Petersen, "Detecting rare events in video using semantic primitives with HMM," in *Proc. of ICPR*, 2004, pp. IV: 150–154.
11. C. Stauffer and W. Grimson, "Learning patterns of activity using real-time tracking," *Pattern Analysis and Machine Intelligence, IEEE Transactions on*, vol. 22, no. 8, pp. 747–757, 2000.

12. H. Zhong, J. Shi, and M. Visontai, "Detecting unusual activity in video," in *Proc. of CVPR*, 2004, pp. 819–826.
13. D. Zhang, D. G. Perez, S. Bengio, and I. McCowan, "Semi-supervised adapted HMMs for unusual event detection," in *Proc. of CVPR*, 2005, pp. I: 611–618.
14. T. Graepel, K. Lauter, and M. Naehrig, "ML Confidential: Machine Learning on Encrypted Data," in *Proc. of ICISC*, vol. 7839, 2012, pp. 1–21.
15. Z. Brakerski, C. Gentry, and V. Vaikuntanathan, "(leveled) fully homomorphic encryption without bootstrapping," in *Proc. of ITCS*, 2012, pp. 309–325.
16. X. Liu, R. Lu, J. Ma, L. Chen, and B. Qin, "Privacy-Preserving Patient-Centric Clinical Decision Support System on Naive Bayesian Classification," *Biomedical and Health Informatics, IEEE Journal of*, to appear.
17. B. Samanthula, Y. Elmehdwi, and W. Jiang, "k-Nearest Neighbor Classification over Semantically Secure Encrypted Relational Data," *Knowledge and Data Engineering, IEEE Transactions on*, vol. 27, no. 5, pp. 1261–1273, May 2015.
18. Paillier and Pointcheval, "Efficient public-key cryptosystems provably secure against active adversaries," in *Proc. ASIACRYPT: Advances in Cryptology*, 1999.
19. M. Barni, P. Failla, R. Lazzeretti, A. Sadeghi, and T. Schneider, "Privacy-preserving ecg classification with branching programs and neural networks," *Information Forensics and Security, IEEE Transactions on*, vol. 6, no. 2, pp. 452–468, June 2011.
20. R. Bost, R. Popa, S. Tu, and S. Goldwasser, "Machine Learning Classification over Encrypted Data," in *Proc. of NDSS*, 2015, pp. 1–14.
21. J. Yuan and S. Yu, "Privacy Preserving Back-Propagation Neural Network Learning Made Practical with Cloud Computing," *IEEE Transactions on Parallel and Distributed Systems*, vol. 25, no. 1, pp. 212–221, 2014.
22. Boneh, Goh, and Nissim, "Evaluating 2-DNF formulas on ciphertexts," in *Theory of Cryptography Conference (TCC), LNCS*, vol. 2, 2005.
23. P. Fong and J. Weber-Jahnke, "Privacy Preserving Decision Tree Learning Using Unrealized Data Sets," *IEEE Transactions on Knowledge and Data Engineering*, vol. 24, no. 2, pp. 353–364, 2012.
24. S. Goldwasser and S. Micali, "Probabilistic Encryption and How to Play Mental Poker Keeping Secret All Partial Information," in *Proc. STOC*, 1982, pp. 365–377.
25. C. Gentry, "Fully homomorphic encryption using ideal lattices," in *Proc. of STOC*, 2009, pp. 169–178.
26. C. Gentry and S. Halevi, "Implementing Gentry's fully-homomorphic encryption scheme," in *Proc. of EUROCRYPT*, 2011, pp. 129–148.
27. I. Damgard, M. Geisler, and M. Kroigard, "Homomorphic encryption and secure comparison," *International Journal of Applied Cryptography*, vol. 1, pp. 22–31, 2008.
28. R. Rivest, "Cryptography and machine learning," *Lecture Notes in Computer Science*, vol. 739, pp. 427–439, 1993.
29. T. T. A. Dinh and A. Datta, "Stream on the sky: Outsourcing access control enforcement for stream data to the cloud," *CoRR*, vol. abs/1210.0660, 2012.
30. J. Bernardo and A. Smith, *Bayesian Theory*. Wiley, 2000.

Chapter 6
Access Control for MHN

In this chapter, we present access control schemes in MHNs. We mainly introduce several promising attribute based access control schemes, associated with some basic attribute-based encryption schemes such as CP-ABE, KP-ABE and multi-authority ABE.

6.1 Introduction

MHNs offer diverse applications for users and data requestors for healthcare and other services, such as health information sharing and exchanging. Among these applications, MHNs allow a data owner (e.g., patient, user) to create, manage, and control his personal health-related data in a cloud server. Other data requestors can access these health data in an owner-defined way from anywhere and at any time. Meanwhile, data owners have the full control of their remotely stored health data which can be efficiently shared and accessed by authorized entities. For example, the patient's heart rate and blood pressure are continuously measured by wearable devices and sent to the cloud server to release the local storage burden. This patient should be able to define the health data access policy which enables other entities to access his data. A neurologist can access the patient's data efficiently to observe the patient's real-time condition. An insurance company representative should not be able to access these real-time health data as shown in Fig. 6.1. Therefore, the health data access policy of MHNs should be defined and used to authenticate the user's identity and authorized access.

Some traditional cryptograph schemes, such as Identity Based Access Control (IBAC) and Role Based Access Control (RBAC) can guarantee the basic access control requirements [1]. IBAC adopts users' identities as a foundation for access control. The cloud server maintains a so-called Access Control Lists (ACLs), which

© Springer International Publishing Switzerland 2015
K. Zhang, X. Shen, *Security and Privacy for Mobile
Healthcare Networks*, Wireless Networks, DOI 10.1007/978-3-319-24717-5_6

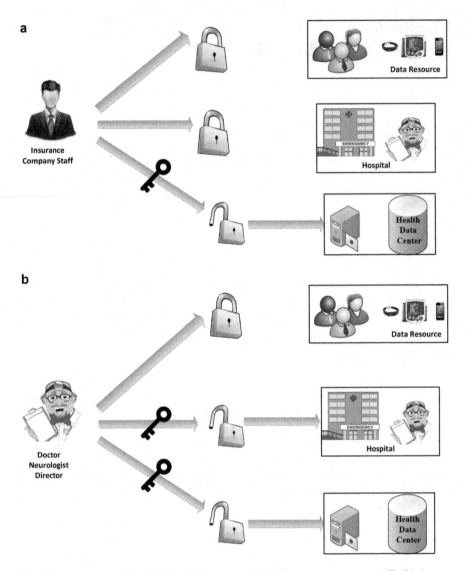

Fig. 6.1 Access control in MHNs. (**a**) Access control of insurance company staff. (**b**) Access control of doctor

include data and provide a list of data requestors with various access rights to the data. It is difficult to maintain such an ACL in a long period as user's access policy may change with the time. For example, when a patient changes hospital, the access policy should be re-defined since the identities are all changed. It is possible to address this issue by using RBAC, where a role represents a bundle of privileges. In RBAC, data requestors are assigned with different roles, such as family doctors, neurologist, insurance company, which authorize privileges of access to the certain data from data owners. For example, a new doctor only needs to be assigned with the role of "doctor" to obtain the access authorization to the health data from data owners. However, RBAC requires deliberate design and has drawbacks if entities' roles change. For example, when new data come into the cloud server, new access roles should be defined.

Attribute-Based Encryption (ABE) scheme is a promising tool to support user-defined access control in MHNs [2–4]. ABE is first proposed introduced by Sahai and Waters [5] in 2005. It is a cryptographic primitive with the PKE where the messages can be encrypted and decrypted based on the basis of users' attributes. A certain ciphertext can be decrypted if and only if the attributes and the decryption keys are matched. ABE schemes can allow users to selectively share the encrypted data and also be extended to provide a fine-grained access [6, 7]. Therefore, Attribute-Based Access Control (ABAC) emerges as such an efficient access control strategy where the access control decisions are based on the attributes of principals and resources.

In this chapter, we investigate ABAC for MHNs. We first present ABE, which is the preliminary of ABAC. We present some typical ABE schemes, such as CP-ABE, KP-ABE and multi-authority based ABE. We also provide some emerging ABAC schemes for health data access control in MHNs.

The remainders of this chapter are organized as follows. Section 6.2 introduces several typical ABE schemes, i.e., CP-ABE, KP-ABE, and multi-authority based ABE. In Sect. 6.3, we present ABAC for MHNs and review some promising access control solutions for e-healthcare system and MHN applications. Finally, we conclude this chapter in Sect. 6.4.

6.2 Preliminaries

A typical implementation method of ABAC is to use ABE techniques. We present several important ABE techniques in the following sections, i.e., CP-ABE, KP-ABE and multi-authority based ABE.

6.2.1 System Model and Attacker Model

In Fig. 6.2, the ABE system consists of a trusted authority (TA), cloud server, data owners and data requestors.

TA is a global trusted authority, initializing the system and assigning key materials for users. Then, TA is not involved in the attribute management and policy making phases. Each data owner packs his health data according to the defined attributes. Then, data owners define their access policy and produce access tree structure for ABE. Data are finally sent to the cloud server in ciphertext. The cloud

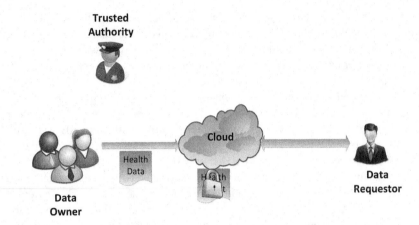

Fig. 6.2 ABE system model

server helps data owner to control the access from data requestors. Data requestors with a set of attributes can receive their secret keys used to decrypt the ciphertext from the cloud server.

In the ABE system, TA is fully trusted without any collusion with other entities. The cloud server is honest-but-curious about the content of the ciphertext. It can honestly follow the protocols. Data owners and requestors may be interested in other entities' attributes and perform collusion with each other.

6.2.2 CP-ABE Implementation

Ciphertext Policy Attribute-Based Encryption (CP-ABE) [8] allows a message encrypted under a user defined access policy. Data owner's secret keys are associated with a set of defined attributes. As shown in Fig. 6.3, CP-ABE reverses the role of encryption and key derivation. The encrypted ciphertext is associated with a user-defined access structure T which is constructed using a certain access policy. If the decryption users can satisfy the access structure defined by data owner, they can successfully decrypt the message.

CP-ABE relies on a user defined access structure. Formally speaking, let $\{P_1, \cdots, P_n\}$ be a set of parties. A collection $\mathbb{A} \subseteq 2^{\{P_1, \cdots, P_n\}}$ is monotone if $\forall B, C$ follow that if $B \subseteq C$ and $B \in \mathbb{A}$ then $C \in \mathbb{A}$. A (monotone) access structure is a (monotone) collection $\mathbb{A}$ of non-empty subsets of $\{P_1, \cdots, P_n\}$. $\mathbb{A} \subseteq 2^{\{P_1, \cdots, P_n\}} \setminus \{\emptyset\}$. The sets in $\mathbb{A}$ are authorized sets, while the ones outside $\mathbb{A}$ are unauthorized sets.

A preliminary of CP-ABE is an access tree $\mathbb{T}$ (with the root r) where each non-leaf node represents a threshold gate. It is described by its children and a threshold value. Let N_x be the number of children of a node x and k_x be the threshold value. $0 \leqslant k_x \leqslant N_x$. The threshold gate is the OR gate if $k_x = 1$; it is the AND gate if

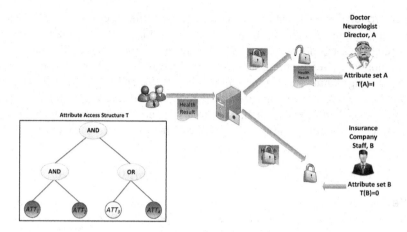

Fig. 6.3 CP-ABE implementation

$k_x = N_x$. Every leaf node y represents an attribute and the threshold $k_y = 1$. Denote a function $\mathsf{parent}(x)$ as the parent of node x. $\mathsf{att}(x)$ denotes the attribute associated with the leaf node x only if x is a leaf node. $\mathbb{T}$ defines an ordering between the children of every node where the children of a node are numbered from 1 to *num*. Another function $\mathsf{index}(x)$ outputs a number associated with the node x. This unique index value is randomly assigned to each node for a given key. In addition, denote $\mathbb{T}_x$ as the sub-tree of $\mathbb{T}$ rooted at node x. $\mathbb{T} = \mathbb{T}_r$. $\mathbb{T}_x(\gamma) = 1$ when a set of attributes γ satisfy the access tree $\mathbb{T}_x$. $\mathbb{T}_x(\gamma)$ can be recursively calculated in the following steps. If x is a non-leaf node, $\mathbb{T}_{ch}(\gamma)$ is computed for all children ch of x. $\mathbb{T}_x(\gamma) = 1$ when at least k_x children output 1. When x is a leaf node, $\mathbb{T}_x(\gamma) = 1$ if $\mathsf{att}(x) \in \gamma$.

The lagrange coefficient is also adopted in CP-ABE. It is defined as $\triangle_{i,S}$ where $i \in \mathbb{Z}_q$ and S is a set of elements in $\mathbb{Z}_q$.

$$\triangle_{i,S}(x) = \prod_{j \in S, j \neq i} \frac{x-j}{i-j}.$$

A CP-ABE scheme consists of four algorithms, i.e., SetUp, Encryption, KeyGen, Decryption. Let $\mathbb{G}_1$ and $\mathbb{G}_2$ be two multiplicative cyclic groups with the same prime order p, and g is the generator of $\mathbb{G}_1$. Let e be a bilinear pairing, where $e \colon \mathbb{G}_1 \times \mathbb{G}_1 \to \mathbb{G}_2$ [9] between $\mathbb{G}_1$ and $\mathbb{G}_2$ exists under two conditions: (1) for any random numbers $a, b \in \mathbb{Z}_q$, $e(g^a, g^b) = e(g,g)^{ab}$; (2) $e(P,P) \neq 1$.

The SetUp algorithm takes as input a security parameter κ, two random numbers $\alpha, \beta \in \mathbb{Z}_q$. $\mathsf{H} \colon \{0,1\}^* \to \mathbb{G}_1$ is a cryptographic hash function. It outputs the public key as

$$\mathsf{PK} = (\mathbb{G}_1, g, h = g^\beta, e(g,g)^\alpha, \mathsf{H})$$

where the master key $\mathsf{MK} = (\beta, g^\alpha)$.

Encryption($\mathsf{PK}, m, \mathbb{T}$) encrypts message m under the access structure $\mathbb{T}$. It first selects a polynomial q_x for each (non-leaf and leaf) node x in $\mathbb{T}$. The polynomial is selected in a top-down manner. It starts from the root node r. The degree d_x of polynomial q_x representing node x in the tree is $k_x - 1$. For root node r, the encryption algorithm selects a random number $s \in \mathbb{Z}_q$ and sets $q_r(0) = s$. Then, it randomly selects d_r other points of the polynomial q_r. For every other node x, it sets $q_x(0) = q_{\text{parent}(x)}(\text{index}(x))$. Then, it randomly selects d_x other points to define q_x.

Denote Y as the set of leaf nodes in $\mathbb{T}$. The ciphertext is generated with the given access tree $\mathbb{T}$ as

$$\mathsf{CT} = (\mathbb{T}, \tilde{C} = m \cdot e(g,g)^{\alpha s}, C = h^s, \forall y \in Y : C_y = gq_y(0), C_y' = \mathsf{H}(\text{att}(y))^{q_y 0})$$

KeyGen(MK, S) takes as input a set of attributes S and outputs a key identifying with this attribute set. It first selects a random number $t \in \mathbb{Z}_q$ and other random number $t_j \in \mathbb{Z}_q$ for every attribute $j \in S$. The key is

$$\mathsf{SK} = (D = g^{\frac{\alpha+\gamma}{\beta}}, \forall j \in S : D_j = g^t \cdot \mathsf{H}(j)^{t_j} \text{ and } D_j' = g^{t_j}).$$

Decryption(SK, CT) includes a recursive algorithm DecNode($\mathsf{SK}, \mathsf{CT}, x$) which takes as input a secret key SK associated with a set of attributes S, the ciphertext $\mathsf{CT} = (\mathbb{T}, \tilde{C}, C, \forall y \in Y : C_y, C_y')$, and a node x in the access tree $\mathbb{T}$.

When x is a leaf node, $i = \text{att}(x)$. If $i \in S$,

$$\mathsf{DecNode}(\mathsf{SK}, \mathsf{CT}, x) = \frac{e(D_i, C_x)}{e(D_i', C_x')}$$

$$= \frac{e(g^t \cdot \mathsf{H}(i)^{t_i}, gq_x(0))}{e(g^{t_i}, \mathsf{H}(i)^{q_x 0})}$$

$$= e(g,g)^{t \cdot q_x(0)}.$$

If $i \notin S$, DecNode($\mathsf{SK}, \mathsf{CT}, x$) $= \perp$.

When x is a non-leaf node, DecNode($\mathsf{SK}, \mathsf{CT}, x$) is run by all node ch that are children of x, and outputs F_{ch}. Let S_x be an arbitrary k_x-sized set of children ch such that $F_{ch} \neq \perp$. If there exists such a set, $F_x = \prod_{ch \in S_x} F_{ch}^{\triangle_{i,S_x'}(0)}$ where $i = \text{index}(x)$ and $S_x' = \{\text{index}(ch) \text{ for } z \in S_x\}$.

$$F_x = \prod_{ch \in S_x} \left(e(g,g)^{t \cdot q_{ch}(0)} \right)^{\triangle_{i,S_x'}(0)}$$

$$= \prod_{ch \in S_x} \left(e(g,g)^{t \cdot q_{\text{parent}(ch)}(\text{index}(ch))} \right)^{\triangle_{i,S_x'}(0)}$$

$$= \prod_{ch \in S_x} e(g,g)^{t \cdot q_x(i) \cdot \triangle_{i,S_x'}(0)}$$

$$= e(g,g)^{t \cdot q_x(0)}.$$

F_x is the output. If no such a set exists, this node does not satisfy and outputs $\perp$.

Decryption(SK, CT) calls DecNode(SK, CT, x) starting from the root node r in the tree $\mathbb{T}$. If the tree is satisfied by S,

$$A = \mathsf{DecNode}(SK, CT, r)$$
$$= e(g,g)^{t \cdot q_r(0)}$$
$$= e(g,g)^{t \cdot s}.$$

Finally, the message m can be decrypted by computing

$$\frac{\tilde{C}}{e(C,D)/A} = \frac{m \cdot e(g,g)^{\alpha s} \cdot e(g,g)^{ts}}{e(h^s, g^{\frac{\alpha+\gamma}{\beta}})} = m.$$

The security of CP-ABE is guaranteed based on Decisional Bilinear Diffie-Hellman (DBDH) assumption and Decisional Modified Bilinear Diffie-Hellman (DMBDH) assumption.

Definition 1 (Decisional Bilinear Diffie-Hellman (DBDH) Assumption). Suppose a challenger randomly selects $a, b, c, z \in \mathbb{Z}_q$. The DBDH assumption is that no polynomial-time adversary is able to distinguish $(A = g^a, B = g^b, C = g^c, Z = e(g,g)^{abc})$ from $(A = g^a, B = g^b, C = g^c, Z = e(g,g)^z)$ with more than a negligible advantage.

Definition 2 (Decisional Modified Bilinear Diffie-Hellman (DMBDH) Assumption). Suppose a challenger randomly selects $a, b, c, z \in \mathbb{Z}_q$. The DMBDH assumption is that no polynomial-time adversary is able to distinguish $(A = g^a, B = g^b, C = g^c, Z = e(g,g)^{\frac{ab}{c}})$ from $(A = g^a, B = g^b, C = g^c, Z = e(g,g)^z)$ with more than a negligible advantage.

In addition, Zhou et al. [10] improve CP-ABE with a constant-size ciphertext which significantly reduces the overhead of encryption and decryption of CP-ABE. Besides the encrypted message and access structure in ciphertext, a ciphertext only requires 2 bilinear group elements. However, CP-ABE is limited in terms of specifying the access policies and management of the user attributes [11].

6.2.3 KP-ABE Implementation

To achieve fine-grained access control, the access structure should be specified in the secret key, and the ciphertexts are labeled associated with a set of descriptive attributes. Goyal et al. [6] propose Key Policy Attribute-Based Encryption (KP-ABE) to implement fine-grained access control. Different from CP-ABE, KP-ABE includes the access policies associated with the secret keys where the ciphertext is labeled by a set of descriptive attributes. The ciphertext can only be decrypted if the

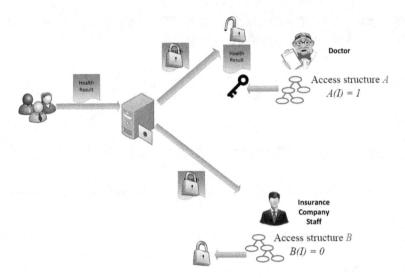

Fig. 6.4 KP-ABE

data attributes satisfy the certain access structure defined by users (i.e., data owners). However, the KP-ABE has limited capabilities to allow the encryptor to determine the decryptor of the ciphertext [6].

Figure 6.4 shows a KP-ABE scheme. The data owner encrypts a message by using a set of attributes. He also defines an access structure, which is a threshold tree of the policy enforced by the data owner. The decryption users first check if their attributes satisfy the access structure defined by the data owner. If so, this decryption user is able to derive the decryption key and recover the message. Otherwise, the decryption key cannot be derived without a satisfied attribute structure. Therefore, the key idea of KP-ABE is that the decryption key is associated with the access structure defined by data owners. Similar to CP-ABE, the security of KP-ABE also replies on DBDH assumption and DMBDH assumption [5].

6.2.4 Multi-authority ABE

In the traditional single authority ABE system, users have to register to a trusted authority and prove their own identities for secret key that helps users to decrypt messages. It is also essential for the individual user to prove that he has a certain set of valid attributes, which allow him to receive secret keys corresponding to each possessed attributes. In such a case, a trusted authority monitoring all attributes should exist in the ABE system. However, it is necessary to have multiple authorities operating simultaneously each of which generates a different set of attributes.

Multi-Authority Attribute-Based Encryption (MA-ABE) allows multiple trusted authorities to collectively generate user's keys. These trusted authorities are responsible to manage the subsets of the attributes of data owners. The data requestor can obtain a part of the secret key from each trusted authority, which can resist collusion attacks [12, 13].

6.3 Attribute Based Access Control

ABAC is contrasted with the sophisticated IBAC and RBAC. ABAC allows the decryption keys to be directly derived from the attributes of data requestors by using the data owner defined access policy. For example, the health data that can be accessible to doctors are marked as such by indicating that it is available to the data requestors that have the doctor attribute. A boolean gate on the attributes, such as "Neurologist" OR ("Doctor" AND "Hospital") can be used to represent the access policy. Note that the attributes are boolean variables indicating whether a user has that attribute or not. In addition, fine-grained attributes (assigned in hierarchical attribute spaces) is necessary in the complicated access policy of MHNs. The user-centric access policies can be achieved by using non-trivial integration of multiple attributes. ABAC can be also effectively combined with IBAC and RBAC in various ways, such as using attributes to populate an ACL or assign a role.

Besides the general access control policies, it is also critical to ensure the fine-grained access in accordance to users' attributes. In MHNs, the dynamic access management is necessary to address the issue of users' attribute changing, revocation [14], new user's participation, etc. In addition, the overheads for different access levels should be balanced to release the computation burden for users. In terms of performance, ABE schemes usually consumes a lot of overhead during the decryption phase since it takes some bilinear computation steps [15].

To achieve fine-grained access, confidentiality and scalability of the encrypted health data [16], the encrypted data from a data owner are shared with multiple users by distributing the keys. Data owners should be able to delegate the computational tasks to the untrusted cloud server, where the attribute based access policy is enforced. This scheme also achieves the accountability of the users' secret keys. The re-encryption of health data and update of the secrete keys are also delegated to the cloud server. To reduce the heavy computation overheads caused by re-encryption of data and update of secret key, re-encryption techniques are integrated with the KP-ABE technique. It is necessary to restrict the revoked users from learning the updated data and keys if the data are modified after user revocation.

In the recent decade, many research efforts have been put to access control for e-healthcare system and MHN applications. Ruj et al. [17] propose an ABAC scheme to achieve anonymity [18] of the data owners. Even though the data owner's identity is hidden, his credentials are still able to be verified. This scheme can resist the replay attacks and accomplish the distribution of the keys in a decentralized pattern. In [19], Hupperich et al. adopt ABE to achieve the confidentiality of the health data. ABE and PKE are integrated for scalable authorization secret keys. Each patient has a smart card associated with a secret PIN, which can be used for the authentication and authorization by using PKE. The patient is provided with a so-called "Transaction Access Code" (TAC) that may be sent to a physician via a smartphone or through any other channel. The physician in turn creates the patient's health data. After the TAC is verified, the health data are encrypted and sent to the cloud server for storage. The decryption requires the data requestors to obtain the TAC and acquire authentication from a Private Key Generator.

Broadcast ciphertext-policy Attribute-Based Encryption (bABE) is an effective scheme to directly revoke the user's keys without either refreshing system parameters or data re-encryption. Although the bABE achieves the health data confidentiality, it also causes an increasing computational overheads in enforcing the access policies. In [20], Nararyan et al. propose another ABAC platform for e-healthcare system where the patients encrypt their health data by using bABE. The health data are accessible to the users satisfying the access policy associated with the ciphertext, with an additional functionality, i.e., user revocation. This platform addresses the key management problem through the users' attributes for data encryption and allowing every user to have only one secret key for the attribute set. It also enables privacy-preserving keyword search over the encrypted health data without disclosing the partial matches or keywords to the untrusted cloud server. Furthermore, the platform permits the healthcare service providers (e.g., hospitals) to perform keyword-based search on the patients' health data. The keyword search is achieved by integrating the bABE and the Public-Key Encryption with Keyword Search (PEKS) [21].

Li et al. [22] propose an ABAC scheme for health data with multi-owner in the multi-authority and multi-user cloud environment. As the data owners can determine their preferences and generate the decryption keys by using the MA-ABE, the decryption keys can be subsequently distributed to the authorized data requestors. Moreover, each authority manages the attributes in a distributed pattern. To reduce the complexity of key distribution, this scheme [22] consists of several security domains. Each security domain is responsible to manage a set with only limited number of users. It is flexible and can support on-demand and efficient revocation of the user's access privileges. However, the it suffers from the excessive computational overhead at the data owner side. To improve the efficiency, Li et al. [23] propose a secure health data sharing framework with multi-owners, which is similar to [22]. The ABE is also adopted to encrypt the data owner's health data. This scheme reduces the costs of key management for data requestors and data owners, and preserves their private attribute from disclosing. It also achieves efficient management and on-demand user/attribute revocation. Although it enhances the scalability of the ABAC in MHNs, the efficiency is still a big challenge when dealing with the situations where data access privileges are granted according to the users' identities other than attributes.

In addition, Jung et al. [24] propose a semi-anonymous privilege control (Anony-Control) scheme to achieve data privacy and user identity privacy. To protect the identity leakage, AnonyControl decentralizes the traditional central authority and achieves semi-anonymity. Furthermore, AnonyControl can generalize data access control to the privilege control, where the privileges of all operations on the cloud data can be fine-grained controlled. Finally, an enhanced fully prevents the identity leakage and achieve the full anonymity. To grant access to the health data according to access privileges, an efficient and secure patient-centric access control (ESPAC) scheme is proposed for the cloud storage by using CP-ABE [25]. The ESPAC utilizes IBE for secure health data transmission between the cloud server and remote patients. The functionality of access control is implemented with CP-ABE. The ESPAC scheme is also able to resist the DoS attacks in a dual server mode. However, it still faces the access flexibility and dynamicity, which results in an inefficient health data transmission to the hospital servers. Alshehri et al. [26] also utilize CP-ABE to encrypt the health data based on the healthcare service providers' credentials and attributes in the cloud environment with multiple data owners. The healthcare service providers share one public key for the health data encryption to reduce the public key distribution and management overheads. Akinyele et al. [27] achieve a flexible secure encryption of patient's health data when health data are transmitted outside the trust boundaries of the healthcare organization, i.e., hospitals. A so-called policy engine produces diverse access policies over these health data according to the user types, such as patient, doctor, physician and insurance company agent. Furthermore, the policy engine identifies a set of attributes including patient age, record type and date/time to encrypt the data by using CP-ABE.

Although ABAC is widely applied in cloud storage system and healthcare system, it still faces challenges when developing MHNs. The high computational overheads of ABAC, especially during revocation, hinder the flourish of ABAC in MHNs.

6.4 Summary

In this chapter, we have investigated access control in MHNs, especially attribute based access control. We have reviewed the basic ABE schemes, i.e., CP-ABE, KP-ABE and multi-authority ABE. With these preliminaries, we have also presented some ABAC schemes in MHNs and health-related applications.

References

1. M. Barua, R. Lu, and X. Shen, "SPS: Secure personal health information sharing with patient-centric access control in cloud computing," in *Proc. of IEEE GLOBECOM*, 2013, pp. 647–652.
2. X. Liang, R. Lu, L. Chen, X. Lin, and X. Shen, "PEC: A privacy-preserving emergency call scheme for mobile healthcare social networks," *Journal of Communications and Networks*, vol. 13, no. 2, pp. 102–112, 2011.

3. X. Liang, M. Barua, L. Chen, R. Lu, X. Shen, X. Li, and H. Luo, "Enabling pervasive healthcare through continuous remote health monitoring," *IEEE Wireless Communications*, vol. 19, no. 6, pp. 10–18, 2012.

4. K. Zhang, K. Yang, X. Liang, Z. Su, X. Shen, and H. Luo, "Security and privacy for mobile healthcare networks — from quality-of-protection perspective," *IEEE Wireless Communications*, vol. 22, no. 4, pp. 104–112, 2015.

5. Sahai and Waters, "Fuzzy identity-based encryption," in *Proc. of EUROCRYPT*, 2005, pp. 457–473.

6. V. Goyal, O. Pandey, A. Sahai, and B. Waters, "Attribute-based encryption for fine-grained access control of encrypted data," in *Proc. of CCS*, 2006, pp. 89–98.

7. K. Zhang, X. Liang, R. Lu, and X. Shen, "Exploiting multimedia services in mobile social network from security and privacy perspectives," *IEEE Communications Magazine*, vol. 52, no. 3, pp. 58–65, 2014.

8. J. Bethencourt, A. Sahai, and B. Waters, "Ciphertext-Policy Attribute-Based Encryption," in *Proc. of IEEE S & P*, 2007, pp. 321–334.

9. D. Boneh, C. Gentry, B. Lynn, and H. Shacham, "Aggregate and verifiably encrypted signatures from bilinear maps," in *Proc. of EUROCRYPT*, 2003.

10. Z. Zhou, D. Huang, and Z. Wang, "Efficient Privacy-Preserving Ciphertext-Policy Attribute Based-Encryption and Broadcast Encryption," *IEEE Transactions on Computers*, vol. 64, no. 1, pp. 126–138, 2015.

11. A. Abbas and S. U. Khan, "A Review on the State-of-the-Art Privacy-Preserving Approaches in the e-Health Clouds," *IEEE Journal of Biomedical and Health Informatics*, vol. 18, no. 4, pp. 1431–1441, 2014.

12. M. Chase and S. Chow, "Improving privacy and security in multi-authority attribute-based encryption," in *Proc. of CCS*, 2009, pp. 121–130.

13. K. Yang, X. Jia, K. Ren, and B. Zhang, "DAC-MACS: Effective data access control for multi-authority cloud storage systems," in *Proc. of IEEE INFOCOM*, 2013, pp. 2895–2903.

14. S. Yu, C. Wang, K. Ren, and W. Lou, "Attribute based data sharing with attribute revocation," in *Proc. of ASIACCS*, 2010, pp. 261–270.

15. H. Lin, J. Shao, C. Zhang, and Y. Fang, "CAM: Cloud-Assisted Privacy Preserving Mobile Health Monitoring," *IEEE Transactions on Information Forensics and Security*, vol. 8, no. 6, pp. 985–997, 2013.

16. S. Yu, C. Wang, K. Ren, and W. Lou, "Achieving Secure, Scalable, and Fine-grained Data Access Control in Cloud Computing," in *Proc. of IEEE INFOCOM*, 2010, pp. 534–542.

17. S. Ruj, M. Stojmenovic, and A. Nayak, "Privacy preserving access control with authentication for securing data in clouds," in *Proc. of IEEE CCGRID*, 2012, pp. 556–563.

18. R. Lu, X. Lin, H. Zhu, P. Ho, and X. Shen, "A novel anonymous mutual authentication protocol with provable link-layer location privacy," *IEEE Transactions on Vehicular Technology*, vol. 58, no. 3, pp. 1454–1466, 2009.

19. T. Hupperich, H. Löhr, A.-R. Sadeghi, and M. Winandy, "Flexible patient-controlled security for electronic health records," in *Proc. of IHI*, 2012, pp. 727–732.

20. S. Narayan, M. Gagné, and R. Safavi-Naini, "Privacy preserving EHR system using attribute-based infrastructure," in *Proc. of CCSW*, 2010, pp. 47–52.

21. Boneh, D. Crescenzo, Ostrovsky, and Persiano, "Public key encryption with keyword search," in *Proc. of EUROCRYPT*, 2004, pp. 506–522.

22. M. Li, S. Yu, K. Ren, and W. Lou, "Securing Personal Health Records in Cloud Computing: Patient-Centric and Fine-Grained Data Access Control in Multi-owner Settings," in *Proc. of SecureComm*, vol. 50, 2010, pp. 89–106.

23. M. Li, S. Yu, Y. Zheng, K. Ren, and W. Lou, "Scalable and Secure Sharing of Personal Health Records in Cloud Computing Using Attribute-Based Encryption," *IEEE Trans. Parallel Distrib. Syst*, vol. 24, no. 1, pp. 131–143, 2013.

24. T. Jung, X.-Y. Li, Z. Wan, and M. Wan, "Control cloud data access privilege and anonymity with fully anonymous attribute-based encryption," *Information Forensics and Security, IEEE Transactions on*, vol. 10, no. 1, pp. 190–199, 2015.

25. M. Barua, X. Liang, R. Lu, and X. Shen, "ESPAC: Enabling security and patient-centric access control for ehealth in cloud computing," *International Journal of Security Networks*, vol. 6, no. 2/3, pp. 67–76, 2011.
26. S. Alshehri, S. P. Radziszowski, and R. Raj, "Secure Access for Healthcare Data in the Cloud Using Ciphertext-Policy Attribute-Based Encryption," in *Proc. of IEEE ICDE*, 2012, pp. 143–146.
27. J. Akinyele, C. Lehmann, M. Green, M. Pagano, Z. Peterson, and A. Rubin, "Self-Protecting Electronic Medical Records Using Attribute-Based Encryption," in *Proc. of SPSM*, 2011, pp. 75–86.

Chapter 7
Summary and Future Research Directions

In this chapter, we summarize the highlights of the monograph and discuss several potential research topics for future work.

7.1 Summary

In this monograph, we have investigated security and privacy for MHNs. Based on the analysis and discussion provided throughout this monograph, we present the following highlights.

- We have introduced the MHN architecture and general applications. We have also provided general security and privacy requirements for MHNs. In addition, we have presented some emerging MHN applications associated with the challenging security and privacy issues, i.e., secure health data collection, misbehaviors during health information sharing, privacy leakage with health data processing and access control.
- To guarantee secure health collection in MHNs, we have proposed a privacy-preserving health data aggregation scheme which utilizes the fixed social spots and the social tie between users and social spots to select the optimal relay. With different health data priorities, the forwarding strategies are adjustable and the corresponding delay requirements can be satisfied with the minimum communication overheads. The security analysis demonstrates that the PHDA can preserve identity and data privacy, while it also resists the forgery attack from inside malicious users and outside attackers. The performance evaluation shows that the PHDA satisfies the delay and delivery ratio requirements for the health data with different priorities, and reduces the communication overheads at the same time.

© Springer International Publishing Switzerland 2015
K. Zhang, X. Shen, *Security and Privacy for Mobile
Healthcare Networks*, Wireless Networks, DOI 10.1007/978-3-319-24717-5_7

- To resist misbehaviors in health data sharing, we have proposed a social-based mobile Sybil detection scheme to detect four levels of Sybil attackers with different attacking capabilities. We have investigated mobile user's pseudonym changing behaviors compared with that performed by Sybil attackers, and utilized contact statistics as the criteria of pseudonym changing for mobile Sybil detection. The security analysis demonstrates that the SMSD can resist four levels of Sybil attackers, while the extensive trace based simulation can validate the detection accuracy of the SMSD. The proposed SMSD scheme is a new paradigm in mobile environments, taking the advantages of powerful storage and computing capabilities in the cloud server, and initiates the trend to distinguish Sybil attackers via mobile user's contacts and pseudonym changing. The semi-supervised learning with HMM can offer accurate detection with reasonable training overhead.
- To prevent user's privacy leakage during health data analysis, we have investigated privacy-preserving health data processing schemes. With the basic classification models and homomorphic encryption techniques as preliminaries, we have provided the integration of classification schemes and privacy-preserving computation blocks. We have also presented an MHN application, i.e., patient-centric clinic decision support system with privacy preservation.
- To achieve access control in MHNs, we have investigated access control schemes, especially attribute based access control. We have introduced the basic ABE schemes, including CP-ABE, KP-ABE, and multi-authority ABE. We have also presented some access control schemes in e-healthcare systems and MHN applications.

7.2 Future Research Directions

This monograph introduces the MHN architecture and applications, proposes security and privacy challenges in MHNs and presents some promising solutions to achieve security and privacy goals in MHNs. Although some preliminary results on security and privacy in MHNs are provided, such as privacy-preserving health data aggregation, mobile social-based misbehavior detection, privacy-preserving health data processing, etc., there are still a set of immature security and privacy solutions for MHNs in several directions as follows.

7.2.1 Lightweight and Secure Health Data Encryption

Although the current wearable devices can offer diverse functionalities to real-timely measure many physiology parameters, they consume certain computation and communication overheads, especially when applying security protections with them [1]. Due to the low-power constraints and portability of these wearable devices, the traditional cryptographic schemes may considerably increase the computation and communication overheads. Therefore, it is necessary to develop lightweight cryptograph schemes for health data encryption [2].

Compressive sensing [3] is a prestigious approach to integrating the lightweight data sensing and security (e.g., encryption and signature) from the perspective of QoP [4]. Conventional sampling schemes follow Shannon's celebrated theorem: the sampling rate must be at least twice the maximum frequency present in the signal (i.e., Nyquist rate) [5]. Against the common wisdom in data sampling and acquisition, compressive sensing can recover certain signals (or images) based on fewer measurements and samples than traditional methods. It relies on two principles: (1) sparsity (which pertains to the signals of interest) and (2) incoherence (which pertains to the sensing modality). Having the sensing matrix, the raw data, which can be sparsely expressed in some domain (e.g., time, frequency, or wavelet), are compressed with different rates. During the construction of sensing matrix, it is difficult to find such a matrix with low coefficient between any two columns.

In addition, some emerging cryptograph schemes, such as NTRU [6, 7], can also provide some benefits for health data encryption. The encryption keys of NTRU are easily created with a reasonable key length. The encryption and decryption of NTRU require low memory but perform fast. NTRU has the smallest average power consumption, but the largest message size. Therefore, it is still an open problem to develop lightweight security protection schemes for health data in MHNs and requires further research efforts.

7.2.2 Misbehavior Detection

In social applications of MHN, as the smarter attackers trend to mimic normal users to hide themselves against the security solutions [8, 9], the traditional approaches focusing on resisting the attacking behaviors may not be effective. The misbehavior detection relies on the learning procedures where learning and training are alternatively applied. Furthermore, human intelligence is highly desirable during the misbehavior modeling and detection to adjust the tunable security and privacy solutions.

Crowdsourcing can be also adopted to facilitate the current misbehavior detection [10]. Especially in MHNs, mobile user's detection capability is not as powerful as that at the server side, or even weaker than online users. Outsourcing the detection tasks to the crowd becomes a possible approach to effectively detect misbehaviors. The crowdsourcing users may detect the suspicious Sybil attackers in the early stage via cryptographic schemes, such as authentication of identities associated with user's contacts, event signatures, etc. The collected detection results from these crowdsourcing users would assist to user behavior learning, social graph or community detection, and make the global decision, etc. Therefore, the crowdsourcing based Sybil detection will become a promising tendency for future research directions.

7.2.3 Health Data Processing with Privacy Preservation

During the health data processing in MHNs, it is urgent to allow the cloud server to perform complicated operations on the encrypted data [11, 12]. For example, machine learning and data mining algorithms [13], such as deep learning should be applied to analyze the physiology parameters and predict disease. The current privacy preservation schemes can support some basic operations. Furthermore, the cryptographic overhead of privacy-preserving machine learning is still too high to be applied in the large scale MHNs.

Meanwhile, the anonymity techniques [14, 15] can be integrated with the cryptograph schemes to balance the privacy and the health data usability. The unlinkability is another important feature of privacy-preserving machine learning. However, it can only achieve certain basic security and privacy requirements. From the above discussions, there exist trade offs between the security and complexity of data processing, especially from the perspective of QoP [4].

In addition, side channel attack may be launched to analyze different types of health data or processing results [16, 17]. The traffic flow or communication patterns of MHN users may be analyzed by outside eavesdroppers [18]. For example, a global attacker may spread some malwares in MHNs and monitor the health data flows of users. User's privacy would be violated by such an attacker. In addition, attackers may also analyze the processing data (e.g., processing time and operations used during the processing phase) in the cloud server to infer if the data are critical or not. Then, the identities or roles of patients and doctors may be analyzed by these attackers. Therefore, developing security protections against side channel attack becomes an essential part of future research directions.

7.2.4 Access Control in MHNs

In ABAC for MHNs, revocation is a challenging issue when the number of users or attributes keeps increasing [19]. To address this issue, Yu et al. [20] adopts re-encryption techniques to allow the data owners to delegate tasks of data file re-encryption and user secret key update to cloud servers. Meanwhile, data contents and user's access privilege information are secretly kept without disclosing to any other untrusted entities. However, the computation overheads of revocation become the most important factor in MHN design, since health data are of a large volume and types [21]. As such, the access policy is difficult to define in MHNs and cover all attributes. There exists a trade off between the computation overheads and the number of health data and attributes. Furthermore, how to guarantee the privacy of attribute is another challenging issue when updating user's secret keys. When users join or leave MHNs, the backward secrecy and forward secrecy should also be achieved. Therefore, access control in MHNs still leaves various challenges in the future research.

Finally, we hope this monograph can shed more lights on the security and privacy protection for MHNs. There will be further research efforts along this emerging line.

References

1. D. He, C. Chen, S. Chan, J. Bu, and P. Zhang, "Secure and Lightweight Network Admission and Transmission Protocol for Body Sensor Networks," *IEEE Journal of Biomedical and Health Informatics*, vol. 17, no. 3, pp. 664–674, 2013.
2. K. Zhang, X. Liang, M. Barua, R. Lu, and X. Shen, "PHDA: A priority based health data aggregation with privacy preservation for cloud assisted WBANs," *Information Sciences*, vol. 284, pp. 130–141, 2014.
3. Y. Bao, J. Beck, and H. Li, "Compressive sampling for accelerometer signals in structural health monitoring," *SAGE Publications*, pp. 235–246, 2011.
4. K. Zhang, K. Yang, X. Liang, Z. Su, X. Shen, and H. Luo, "Security and privacy for mobile healthcare networks — from quality-of-protection perspective," *IEEE Wireless Communications*, vol. 22, no. 4, pp. 104–112, 2015.
5. E. J. Candès and M. Wakin, "An introduction to compressive sampling," *IEEE Signal Processing Magazine*, vol. 25, no. 2, pp. 21–30, 2008.
6. J. Hoffstein, J. Pipher, and J. Silverman, "NTRU: A Ring-Based Public Key Cryptosystem," in *Proc. of ANTS*, 1998, pp. 267–288.
7. P. Nguyen and D. Pointcheval, "Analysis and Improvements of NTRU Encryption Paddings," in *Proc. of CRYPTO*, 2002, pp. 210–225.
8. X. Liang, K. Zhang, X. Shen, and X. Lin, "Security and privacy in mobile social networks: challenges and solutions," *IEEE Wireless Communications*, vol. 21, no. 1, pp. 33–41, 2014.
9. K. Yang, K. Zhang, J. Ren, and X. Shen, "Security and Privacy in Mobile Crowdsourcing Networks: Challenges and Opportunities," *IEEE Communications*, vol. 53, no. 8, pp. 75–81, 2015.
10. J. Ren, Y. Zhang, K. Zhang, and X. Shen, "Exploiting Mobile Crowdsourcing for Pervasive Cloud Services: Challenges and Solutions," *IEEE Communications Magazine*, vol. 53, no. 3, pp. 98–105, 2015.
11. X. Liu, R. Lu, J. Ma, L. Chen, and B. Qin, "Privacy-Preserving Patient-Centric Clinical Decision Support System on Naive Bayesian Classification," *Biomedical and Health Informatics, IEEE Journal of*, to appear.
12. T. Graepel, K. Lauter, and M. Naehrig, "ML Confidential: Machine Learning on Encrypted Data," in *Proc. of ICISC*, vol. 7839, 2012, pp. 1–21.
13. R. Bost, R. Popa, S. Tu, and S. Goldwasser, "Machine Learning Classification over Encrypted Data," in *Proc. of NDSS*, 2015, pp. 1–14.
14. K. E. Emam, F. K. Dankar, R. Issa, E. Jonker, D. Amyot, E. Cogo, J.-P. Corriveau, M. Walker, S. Chowdhury, R. Vaillancourt, T. Roffey, and J. Bottomley, "A Globally Optimal k-Anonymity Method for the De-Identification of Health Data," *American Medical Informatics Association*, vol. 5, no. 16, pp. 670–682, 2009.
15. L. Sweeney, "Achieving k-Anonymity Privacy Protection Using Generalization and Suppression," *International Journal of Uncertainty, Fuzziness and Knowledge-Based Systems*, vol. 10, no. 5, pp. 571–588, 2002.
16. S. Chen, R. Wang, X. Wang, and K. Zhang, "Side-Channel Leaks in Web Applications: A Reality Today, a Challenge Tomorrow," in *IEEE Symposium on Security and Privacy*, 2010, pp. 191–206.
17. Forbes. [Online]. Available: http://www.forbes.com/sites/robertvamosi/2015/07/20/side-channel-analysis-can-protect-iot-scada/
18. X. Lin, R. Lu, X. Shen, Y. Nemoto, and N. Kato, "SAGE: A Strong Privacy-preserving Scheme Against Global Eavesdropping for ehealth Systems," *IEEE Journal on Selected Areas in Communications*, vol. 27, no. 4, pp. 365–378, 2009.

19. K. Yang, X. Jia, and K. Ren, "Secure and Verifiable Policy Update Outsourcing for Big Data Access Control in the Cloud," *IEEE Transactions on Parallel and Distributed Systems*, to appear.
20. S. Yu, C. Wang, K. Ren, and W. Lou, "Achieving Secure, Scalable, and Fine-grained Data Access Control in Cloud Computing," in *Proc. of IEEE INFOCOM*, 2010, pp. 534–542.
21. ——, "Attribute based data sharing with attribute revocation," in *Proc. of ASIACCS*, 2010, pp. 261–270.

Printed in the United States
By Bookmasters